An OPUS book

The Making of the Atomic Age

Alwyn McKay

The Making of the Atomic Age

Oxford New York
OXFORD UNIVERSITY PRESS
1984

Oxford University Press, Walton Street, Oxford OX2 6DP
London New York Toronto
Delhi Bombay Calcutta Madras Karachi
Kuala Lumpur Singapore Hong Kong Tokyo
Nairobi Dar es Salaam Cape Town
Melbourne Auckland
and associated companies in
Beirut Berlin Ibadan Mexico City Nicosia

Oxford is a trade mark of Oxford University Press

First published 1984 as an Oxford University Press paperback
and simultaneously in a hardback edition

British Library Cataloguing in Publication Data
McKay, Alwyn
The making of the atomic age. – (OPUS)
1. Atomic power
I. Title
621.48 TK9145
ISBN 0–19–219193–4
ISBN 0–19–289174–X Pbk

Library of Congress Cataloging in Publication Data
McKay, Alwyn.
The making of the atomic age.
(OPUS)
Bibliography: p.
Includes index.
1. Atomic bomb – History. 2. Atomic energy – History.
I. Title. II. Series.
QC773.M35 1984 539.7 84-3931
ISBN 0–19–219193–4
ISBN 0–19–289174–X (pbk.)

Set by Hope Services, Abingdon
Printed in Great Britain by
The Thetford Press Ltd.

To my wife

Preface

This book describes how the atomic age came into being. The period covered is roughly the first sixty years of the present century. The scope is world-wide, involving especially the major belligerents in the Second World War.

The aim has been to report what happened in simple language, covering the whole field in a fairly brief compass, so that the main outlines stand out clearly. Obviously this has necessitated selection both of historical facts and of the names of those involved. The emphasis throughout is on the scientists, and what motivated them. Those who want more information can find it in the books listed under Further Reading.

The initial phase was one of scientific discovery. Then from 1939 onwards, nuclear fission propelled the subject into the military arena, leading to the enormous Manhattan Project in the US that produced the Hiroshima and Nagasaki bombs. Starting in the 1950s came the big industrial development of nuclear power. Non-scientists generally find the later part of the story easier to follow, but the author hopes that they will nevertheless persist with Chapters 1–4, in order to appreciate how the technological side grew out of the science. A brief description of the main features of the atomic nucleus has been included by way of assistance.

The book would be incomplete without some account of the growth of nuclear power, and Chapter 14 gives a short sketch of this very large topic. Controversial aspects cannot be argued fully in so small a space, but the author believes the statements made are a true bill.

The sources of the longer quotations are indicated in the text or at the end of the book. Remarks attributed to individuals are taken from sources believed to give their actual words rather than an imaginative reconstruction.

The views, opinions, and judgements expressed in the book are, of course, those of the author rather than his former employer, the UK Atomic Energy Authority.

He is grateful to his friends for their suggested improvements to the original draft, especially Laura Arnold, Jim and Peter Baynard-Smith, Brian Wade, and Professor Gilbert Walton, and to Angela Rattue for typing the manuscript.

H. A. C. McKAY

Contents

List of Plates

In addition the author is personally indebted to the following for their help in locating and providing suitable photographs: J. Deakin (Cavendish Laboratory, Cambridge), Ms K. Everett (BBC Publications), A. B. Krusciunas (Argonne National Laboratory, US), G. Newman (Oak Ridge National Laboratory, US), Professor Glenn T. Seaborg (Lawrence Radiation Laboratory, US), Ms M. Stewart (Churchill College, Cambridge), Mrs E. Walker (UK Atomic Energy Authority), and R. Wymer (Oak Ridge National Laboratory).

List of Figures

Figure 12 by Taurus Graphics

Figure 17 by Illustra Design

List of Tables

1 A Passion for Atoms

Few people can have known greater happiness than those who discovered the secrets of the atom and its nucleus in the first half of this century. Their work was, to them, absorbing, exciting, and unquestionably important. They were dedicated: they expected little wealth or public recognition – only the thrill of achievement and the acclaim of their colleagues.

What they accomplished was a revolution in our ideas about the nature of matter. Academic? Yes, but it led to the atom bomb and nuclear power.

The nineteenth century had seen the creation of a marvellous edifice of theories to describe the material universe, so elegant and harmonious that we call it 'classical' physics. It had achieved great triumphs such as the prediction of the existence of the planet Neptune and of radio waves, followed in each case by confirmatory observations. Some of the scientists of the time thought that everything important had already been discovered. Yet the next decades were to see an avalanche of new knowledge, much of which existing theories were powerless to explain.

In retrospect it seems surprising that the limitations of classical physics were not better recognized. Most of chemistry, for example, fell outside its scope. The chemists, last century, knew about eighty different kinds of atoms, and they had deduced many of their rules of behaviour, including especially the ways they combine to form molecules. The known laws of physics were not merely incapable of accounting for all this, but appeared almost irrelevant.

In the late 1890s a whole range of hitherto unsuspected phenomena began to come to light. Two discoveries were the result of lucky accidents: those of x-rays by Wilhelm Roentgen in Würzburg in 1895 and of an unusual radiation from uranium, which fogged a photographic plate, by Henri Becquerel in Paris in 1896. Joseph John Thomson's identification in 1897 of the small electrically charged particle called the electron, on the other hand, came out of logically planned investigations into electrical discharges in gases, carried out at the Cavendish Laboratory in Cambridge. Similarly Pierre and Marie Curie's discovery in 1898 of two new chemical elements, polonium and radium, which emitted radiation like that from uranium, but of far greater intensity, was the fruit of systematically following up Becquerel's observations. They named the whole phenomenon 'radioactivity'.

The Curies' protracted struggle, in a cold and ill-equipped shed at the School of Physics in Paris, to confirm the existence of radium by separating it from a whole ton of residues from the uranium mines of Joachimstal, so that it could be seen and measured, is one of the epics of science. For forty-five months they laboured fanatically at their task, living in semi-poverty, neglecting even to feed themselves properly. Their aim was scientific knowledge; nobody yet knew that radium would be any use. Finally they had a tenth of a gram of the hard-won material, a series of research papers to their name, and a growing correspondence with leading scientists around Europe. A year later they found themselves famous, with the ultimate accolade of the Nobel prize, besides other honours. Fame brought them money and a more comfortable existence, but at times they felt it a nuisance, because its demands interfered with their work.

Pierre Curie was the complete absent-minded professor. There is a story that the Curies' cook, fishing for a compliment, asked him if he had enjoyed the steak he had just consumed with gusto. 'Did I eat a steak?' he asked vaguely, and then, sensing that he had said the wrong thing, 'It's quite possible.' He was totally devoted to his work, and took it for granted that Marie too would sacrifice herself to their 'dominating thoughts'. Marie, however, sometimes had hankerings after a more normal life, though she castigated herself for her 'weakness'.

The Institut de Radium in Paris, founded a few years after Pierre's tragic death in a street accident in 1906, was one of their dreams. Marie spent most of her time there after the 1914–18 war, working at her beloved subject of radioactivity.

The Curie saga has become part of the folklore of science, giving people a romantic but far from typical picture of the scientist.

Ernest Rutherford's life and personality provide as great a contrast as could well be imagined. Almost the only thing he had in common with the Curies was a passion for radioactive research. He arrived in Cambridge from New Zealand in 1895 to work with J. J. Thomson, and while there he heard about Becquerel's and the Curies' discoveries. This led him to embark on a lifelong study of radioactivity. From Cambridge he moved to professorships in Montreal (at only 28) and Manchester, and finally he returned to Cambridge where he succeeded J. J. Thomson as Cavendish Professor in 1919. He became Sir Ernest Rutherford in 1914, though his young daughter doubted whether he was dignified enough, and Lord Rutherford of Nelson in 1931. A brilliant career!

At McGill University in Montreal he collaborated with an outstanding young chemist from Oxford, Frederick Soddy, whose skills formed just the right complement to his own. In the course of a bare two years the pair proved by their experiments that the essence of radioactivity is the spontaneous change of one kind of atom into another. This meant abandoning the traditional idea of atoms as tiny unsplittable balls that never change their nature. On the contrary, in radioactivity we observe a transmutation of one chemical element into another – not lead into gold, as the alchemists desired, but for example radium into the inert gas, radon. The radium, as it gradually disappears, is said to undergo radioactive decay.

From then on, Rutherford was in the forefront of the tide of discovery. He became a legendary figure, and numerous anecdotes were told about him, for instance how he would whistle 'Onward, Christian soldiers!' when the work was going swimmingly, but 'Fight the good fight' when in the midst of problems. He was a bluff, vigorous, straightforward man, who might have been taken for a country squire, and he went through life with zest. 'A very jolly man and most stimulating,' said Otto Hahn, who worked with Rutherford in the early Montreal days and who, years later, was to share in the discovery of nuclear fission.

Rutherford had a flair for devising the right experiment with the limited technical resources of the time, and his mind would penetrate to the heart of a problem in a direct, uncomplicated way. There is a famous instance when one of his colleagues showed him the results of firing small electrically charged particles (alpha-particles) at atoms to probe their structure. Very occasionally one of the particles suffered a severe change in direction, sometimes almost bouncing back the way it had come in. Rutherford likened this to a fifteen-inch shell, fired at a piece of tissue paper, rebounding and hitting you. A few of the alpha-particles, he said, must have encountered immensely powerful forces inside the atoms, and this could have happened if the atom had a very small electrically charged nucleus; the majority of the alpha-particles would then pass more or less straight though the empty spaces of

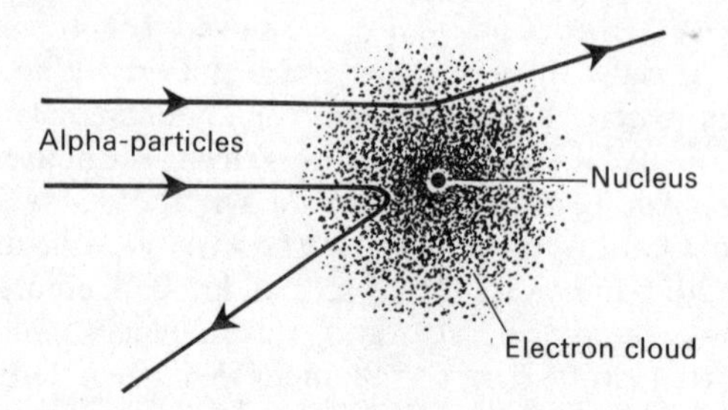

Fig. 1. Paths of alpha-particles through an atom.
An alpha-particle that nearly hits the nucleus is strongly deflected; while others are only slightly deflected.

the atom, but one that chanced to graze the nucleus would be violently deflected (Fig. 1). The idea of the 'nuclear atom' had been born.

Many men who were later to become famous came to work with Rutherford. To name only a few from the continent of Europe, there were Otto Hahn, already mentioned; Hans Geiger, another German, one of the inventors of the Geiger counter; Georg von Hevesy, a Hungarian, who devised many of the techniques of radiochemistry (in which radioactivity is used to investigate chemical problems); and above all, the Dane, Niels Bohr, who became as important a figure as Rutherford himself. As these man in due course went out from Rutherford's laboratory, they established new centres of research into radioactivity and the atomic nucleus. Nearly everyone who worked in the field could indeed trace his 'lineage' back to Rutherford.

During this period there were developments in two other areas of physics that were to prove of great significance in atomic science. The first was Albert Einstein's relativity theory. Among its many consequences is the famous equation relating mass and energy (usually written $E = mc^2$). According to this equation, mass and energy can be regarded as two forms of the same thing – a novel idea at the time – and if we convert a single gram of matter into energy, we shall get as much as from an atom bomb. A very small proportion of this energy is released when one kind of atom changes into another by radioactive decay. A much larger proportion is released in nuclear reactors and nuclear weapons.

The second of these developments also involves a novel idea. Nineteenth-century physicists had been troubled by various anomalies associated with heat and light, where the classical laws gave totally wrong results. These predicted, for example, that a hot piece of iron should give out mainly violet and ultra-violet light, and should not change in colour as the temperature rises. Yet as everyone knows it becomes first red hot, and then in turn orange, white, and blue.

Max Planck explained this in 1900 by introducing the idea that atoms handle energy much as a supermarket now handles butter. Instead of weighing out whatever amount of butter you ask for, a supermarket only sells 250 gram packets. Planck postulated that atoms similarly only accept or give out energy in fixed amounts, which he called 'quanta'. This simple concept explained some at least of the anomalies, giving theoretical laws which fitted some of the experimental results remarkably accurately.

The idea that energy is 'quantized' may seem innocent enough, if a little unusual. Sir James Jeans, the Cambridge astrophysicist, however, has commented that many at the time thought it 'sensational, revolutionary, and even ridiculous'. In fact, the quantum theory, as it is called, marks the turning-point from classical to modern physics.

It is to Niels Bohr, above all, that we owe the revision of physics in the light of this theory. It formed his life's work. He arrived in Cambridge from Copenhagen at the beginning of the autumn term in 1911, just before his twenty-sixth birthday, full of youthful enthusiasm and overjoyed at the prospect of close contact with J. J. Thomson and other famous scientists. Later he spent four months with Rutherford in Manchester where he perceived that classical physics faced a complete breakdown in the atomic domain. He insisted especially that Rutherford's nuclear atom could not survive if it obeyed the laws that work so well for dynamos and electric motors.

In Rutherford's model, the electrons circle the nucleus like the planets round the sun, but there is a difference: the electrons, unlike the planets, carry electric charges. The branch of classical physics known as electrodynamics asserts that in this situation the electrons should radiate energy continuously, and as a result of their loss of energy they should very soon fall into the nucleus. Manifestly this does not happen.

Bohr therefore boldly asserted that classical electrodynamics does not apply to an electron in an atom. He postulated that the atom exists in a 'stationary state' in which the electron orbits the nucleus without giving out energy.

He pondered on this idea for several months. Then in February 1913 a student in Copenhagen drew his attention to certain regularities which had been discovered thirty years before in the spectrum of hydrogen, that is to say, in the colours of the light emitted by hot hydrogen. It was like finding the key piece in a jigsaw puzzle. Bohr now surmised that an atom can exist not just in one, but in a series of stationary states of different energies, and that it can 'jump' from one state to another by absorbing or emitting precisely one quantum of energy. On this basis he was able to derive a mathematical formula which accounted exactly for certain of the lines in the hydrogen spectrum.

Bohr's theory violated the old rule of classical physics that there are no jumps in nature. This was more than some scientists could stomach. Jeans, however, noted that the justification for Bohr's assumptions was 'the very weighty one of success'. Rutherford went to considerable trouble to help Bohr to publicize his ideas, despite his scepticism about theoreticians in general. On a later occasion he said of them, only half in jest, 'They play games with their symbols, but we, in the Cavendish, turn out the real solid facts of nature'. But Bohr, he said, was different.

Among the real solid facts was Rutherford's demonstration in 1919 of the first artificial transmutation of one element into another, as distinct from the natural transmutations of radioactivity. Radioactivity goes its own way very nearly regardless of all attempts to influence it, while the new discovery made it possible to transmute one kind of atom into another in a controllable way.

As in some of his earlier experiments Rutherford bombarded atomic nuclei with alpha-particles from a radioactive preparation. The remarkably simple and small apparatus he used is shown in Plate 4. Previously the interest had been in the deflection of the alpha-particles when they pass close to nuclei; now it was a question of what happens when there is a direct hit. When the nuclei under bombardment were those of nitrogen, Rutherford observed the production of a few particles of a new type, which he was able to identify as the nuclei of hydrogen atoms, known as 'protons'. So – an alpha-particle went into the nitrogen nucleus, and a proton came out. This carried the further highly significant implication that the nitrogen nucleus had been turned into something else, in fact an oxygen nucleus, as the physicists could deduce from a balance-sheet of the electrical charges in the process.

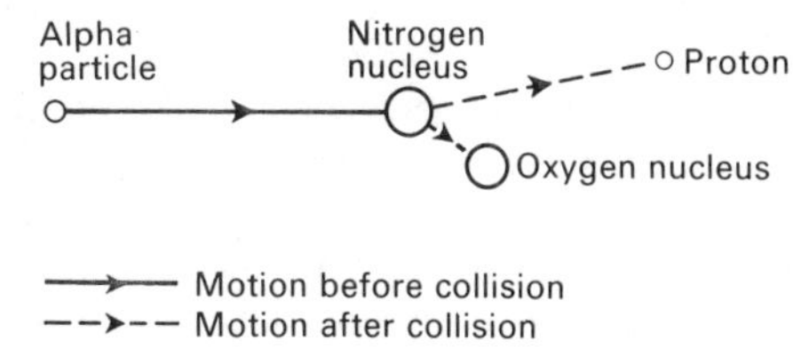

Fig. 2. The first artificial nuclear transmutation.
An alpha-particle striking a nitrogen nucleus transmutes it to oxygen.

Nitrogen had been transmuted into oxygen, though on an exceedingly small scale (Fig. 2).

This discovery was the culmination of Rutherford's time in Manchester. Immediately afterwards he took up his appointment at the Cavendish Laboratory in Cambridge. There, a brilliant group gradually gathered round him, numbering eventually some sixty in all – quite a crowd for those days – and the Cavendish became the nuclear physicist's Mecca.

The first years of Rutherford's reign were relatively quiet, though there was steady progress in areas already opened up. More examples of artificial transmutation, similar to that of nitrogen, were found. Another line was the investigation of atoms which differ only in their nuclei, not in their outer parts, that is to say, the study of isotopes. Isotopes were already known among the radioactive elements, and in one or two other instances. Now with the aid of a special instrument he had invented, the mass

Table 1. Important dates in the early years of nuclear science.

1896	Becquerel observes radiation from uranium, causing fogging of a photographic plate.
1897	Thomson discovers the electron.
1898	The Curies discover the radioactive elements, polonium and radium.
1902	Rutherford and Soddy show that an essential feature of radioactivity is the spontaneous transmutation of one chemical element into another.
1910	The idea of isotopes is established, especially by Soddy.
1911	Rutherford propounds the idea that the atom has a very small, positively charged nucleus.
1913	Bohr propounds his model of the atom.
1919	Rutherford observes the first artificial nuclear transmutation. Aston invents the mass spectrograph for studying isotopes.

spectrograph, Francis Aston showed that nearly all elements are mixtures of isotopes: oxygen has three, chlorine two, and so on. The reason the isotopes had gone undetected for so long is that, because the outer parts of their atoms are virtually identical, almost all aspects of their behaviour are identical too, or very nearly, so that it is difficult to separate them, or even to tell them apart. They do not separate in nature either, and in consequence each element normally consists of a mixture of isotopes in constant, unvarying proportions.

To distinguish between isotopes or to separate them we make use of the properties of their nuclei, which *are* different. The existence of isotopes was indeed first suspected when it was discovered that certain radioactive species differ quite strikingly in their radioactivity, but not in their chemistry. This was explained by the fact that radioactivity is something which happens to the nucleus, while chemistry is concerned almost entirely with the outer parts of the atom. Evidently therefore we can have atoms which differ in their nuclei but not (except to a very minute extent) in their outer parts. In cases where there is no radioactivity to provide a difference, a pair of isotopic atoms still differ in mass: one is heavier than the other. It was this that Aston invoked in his mass spectrograph, which sorted out heavier from lighter atoms, giving a very small-scale separation of the isotopes (Fig. 3).

While the Cavendish probed the nucleus of the atom, physicists

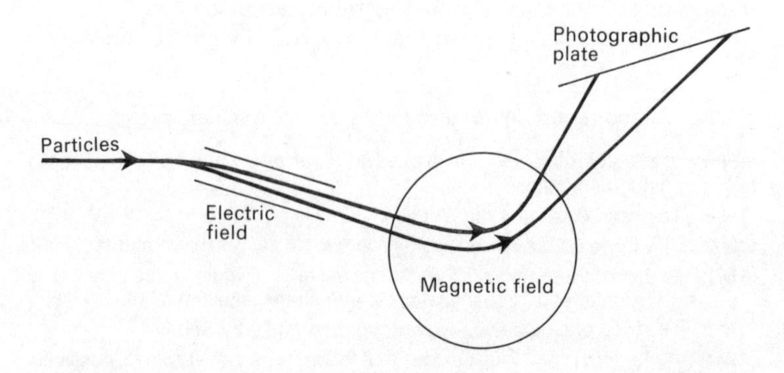

Fig. 3. The principle of the mass spectrograph.

The diagram shows the paths taken by two particles of different mass through the apparatus. Heavier particles are separated from lighter by the time they reach the photographic plate.

elsewhere, especially in Germany, found a happy hunting ground in investigating its outer parts. Here there was scope for powerful mathematical intellects to elaborate Bohr's atomic model, and within remarkably few years the complex secrets of the 'cloud' of electrons round the nucleus had been unravelled. The nucleus seemed a much less tractable theoretical problem, and was left as a puzzle to be solved later.

Not that everything went smoothly in developing the 'Bohr atom'. In the early 1920s the original quantum theory based on Bohr's idea of quantum jumps ran into two kinds of difficulties. On the one hand it sometimes gave the wrong answers (though still tantalizingly nearly right), and on the other it had philosophical shortcomings to which Bohr himself repeatedly drew attention. For instance, in his original atomic model he had thrown out classical electrodynamics, but had used classical mechanics to calculate the energy of an electron in its orbit: why should classical concepts be valid in one case but not the other? Werner Heisenberg, who came from Germany to work with Bohr in 1924, has written that 'the difficulties . . . became more and more embarrassing, [the] internal contradictions seemed to become worse and worse, to force us into a crisis . . .'

Heisenberg himself was the first to find a way out of the crisis. In 1925 while on holiday in Heligoland he had an idea which was soon developed into a radically new theory, known as quantum mechanics. At one blow it removed both the wrong answers and the philosophical difficulties. It gave the physicists a reliable mathematical tool which they could apply to calculations on atomic events.

Quantum mechanics undergirds the whole of the work to be described in this book, and a short account of some of its features is included as an addendum to this chapter, which is not essential for understanding the rest of the book. Yet curiously enough, most of the story of nuclear weapons and nuclear power can be told without mentioning it. Nearly all the major discoveries were made without its direct aid.

However, there was one important exception in the early days of quantum mechanics. The type of radioactivity shown by radium, for example, in which the atomic nucleus throws out an alpha-particle, presented an enigma. The alpha-particle is imprisoned in the nucleus as if by a high wall all round it, and it seems to have only about 20 per cent of the energy necessary to cross the wall. How then does it ever get out?

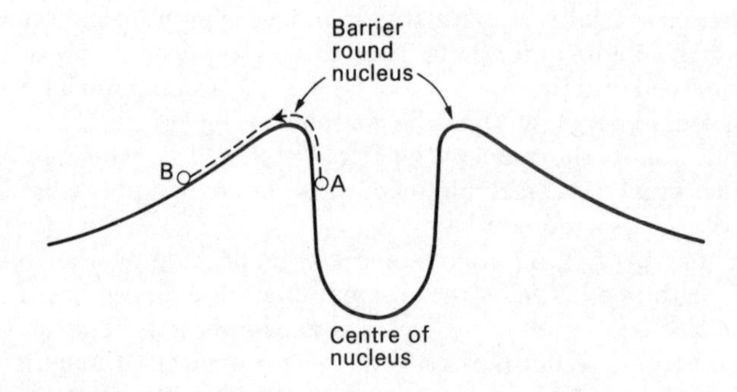

Fig. 4. Escape of an alpha-particle from the nucleus.

According to classical concepts, to get from A (inside the nucleus) to B (outside the nucleus) the alpha-particle must pass over the top of the barrier, as indicated by the dashed line. According to quantum mechanics, it is meaningless to enquire how it gets from A to B. It has a calculable chance of getting to B, even when it has too little energy to reach the top of the barrier.

According to classical physics, escape is impossible. Like a car on a switchback railway, the alpha-particle cannot get to the other side unless it has enough energy to reach the top of the barrier. But quantum mechanics gives a different answer, as George Gamow, a Russian émigré working in Bohr's Institute in Copenhagen, pointed out in 1928. Quantum mechanics is concerned solely with what is actually observed, and regards it as meaningless to enquire what happens between the observations. It proceeds, as it were, in a series of jumps from observation to observation, whereas classical physics supposes a continuous, smooth transition. So quantum mechanics happily envisages the alpha-particle inside the nucleus at one instant and outside it at a later instant, without its passing through any intermediate position, such as the top of the barrier (Fig. 4).

This, at first sight strange, idea is often called the 'tunnel effect', to make it clear that the classical picture of climbing over the barrier has been discarded, though it would be equally false to suppose that the alpha-particle really does tunnel through.

In 1929, Gamow, encouraged by Bohr, paid a visit to the Cavendish. Here his new concept stimulated two young experimentalists, John Cockcroft, destined to become one of the leaders of Britain's nuclear programme, and Ernest Walton.

Cockcroft had come to the Cavendish a couple of years before with a personal letter of introduction to Rutherford. While still at school he had been fascinated by the early discoveries about radioactivity, and before the 1914–18 war he actually went to Rutherford's lectures at Manchester University for 'light relief' from his mathematical studies. After the war, with making a living in mind, he took a course in electrical engineering and worked as an apprentice at Metropolitan-Vickers. But he kept up his interest in nuclear physics, and ultimately headed for Cambridge.

When he and Walton met Gamow there in 1929, they were familiar with the idea of investigating the nucleus by firing high-speed projectiles at it, but hitherto the only projectiles with enough energy had been the alpha-particles from radioactive substances, which Rutherford had used. Could such projectiles be made artificially by accelerating atomic particles to high energies by means of electrical potentials? Perhaps, but to obtain energies like those of alpha-particles, millions of volts would be needed, and this seemed altogether beyond the technical resources of the time. Gamow, however, in a memorandum to Rutherford, pointed out that the tunnel effect indicates a much more hopeful conclusion. Since the particles can get into the nucleus without passing over the top of the nuclear energy barrier, much less energy is necessary, and a potential of a few hundred thousand volts might suffice.

Rutherford therefore authorized Cockcroft and Walton to construct the most expensive piece of equipment yet built at the Cavendish – costing all of £500 – to accelerate protons. Cockcroft's electrical engineering training came into its own, and potentials up to seven hundred thousand volts were obtained – very high for those days. The apparatus is shown in Plate 6, with Walton controlling it from inside a lead-covered tea-chest for radiation protection. The work finally reached fruition in April 1932, when the two experimenters fired their high-speed protons at a target consisting of lithium, chosen because its nuclei have a particularly low energy barrier. Success! Alpha-particles were detected coming out of the target, showing that the protons were indeed penetrating and breaking up the lithium nuclei. A nuclear reaction, a transmutation, had been brought about by entirely artificial means. Moreover the effects observed were a million times as intense as those obtained with alpha-particles.

This was a double triumph – for the quantum theory, which had pointed the way, and for the experimentalists, who had created a new and powerful tool, the particle accelerator or 'atom smasher', as the newspapers liked to call it. The Press rightly presented

Cockcroft and Walton's 'splitting of the atom' as a major breakthrough. Nuclear science was poised for a rapid advance.

A Note on Quantum Mechanics

It is difficult to express quantum mechanics in any other form than as a series of mathematical equations. It offers no simple picture comparable to the 'Bohr atom' with its electrons circulating round the nucleus. This is because our mental images – of atoms as little round balls, for instance – are basically drawn from everyday life, where classical physics reigns. But when we come down to the atomic scale we need new, non-classical concepts, and these are embodied in the equations of quantum mechanics.

The consequences of these equations subvert many of our preconceived ideas. The tunnel effect already mentioned is one example. Another is the quantum-mechanical assertion that Nature herself imposes limits on what we can find out about the universe. For instance, we cannot know both the position and velocity of an electron or any other particle precisely; the more nearly we measure the position, the less accurately do we know the velocity, and vice versa. This is not for lack of experimental skill; however much we improve our technique, the limitation will remain, because it is part of the nature of things. It is referred to as the 'uncertainty principle'.

The principle may seem strange, but it is not actually mysterious, for we cannot locate an object without giving it a slight push. If we use a measuring rod for the purpose and bring it into contact with the object, then to be sure it is really touching, we press it against the object. But this sets the object infinitesimally in motion, so that its velocity is uncertain. The harder we press the measuring rod against the object, the surer we are of the latter's position, but the more doubt there is of its velocity. Even when we say, 'I can see where it is', we do not escape the limitations, because light exerts a minute pressure on anything it strikes. Of course the effects are too small to detect in dealing with pots and pans, but they are important for atoms and electrons.

Equally strange is the conclusion that we cannot foretell the future precisely. Until quantum mechanics came along, scientists from Newton onwards have tended to believe that if only we could accumulate enough information about the present state of the universe, and had an all-powerful computer, science would enable us to predict the future. If, for instance, we knew where an electron was, and how fast and in what direction it was moving, we could calculate where it would be at some future moment. But the uncertainty principle asserts that we cannot know all these things at once. So we are forced to abandon the idea that the universe is predictable in the simple sense of the word as classical physics understood it.

However, the universe is still predictable in another sense. Although we cannot say that an electron will definitely be at point X at a particular moment in the future, we can calculate the *odds* on finding it there, or at some other point Y. This means that if, for example, the odds are fifty-fifty, then if we repeat our experiment a large number of times, we shall find the electron at X in about half the experiments, and at Y in the other half. Quantum mechanics gives us probabilities or statistics rather than unique answers.

Bohr took this to mean that the universe is statistical in its very nature. The information we gather and the predictions we make are subject to uncertainties that can never be eliminated. 'Even God does not know more' was one way of expressing it. Other scientists have, however, been reluctant to accept this view. Einstein had a running argument with Bohr over a period of years, in which he tried to show that there is another layer of reality below the statistical, in which the uncertainties are no longer present. The clash was like a contest between two grandmasters of chess. During a famous scientific conference in 1927 Einstein would produce, over breakfast, a gambit in the form of an imaginary experiment to prove his point, and Bohr would go into a huddle with his colleagues and refute the gambit by teatime.

Bohr approached Einstein's arguments in the expectation of finding subtle fallacies in them. His confidence derived from his philosophy. He had generalized the uncertainty principle into a concept of wide application, which he called 'complementarity'. The uncertainties arise because we cannot observe anything without in some way disturbing it. If we choose one way of observing it, the disturbance we cause excludes another 'complementary' way. We have already seen how this applies in atomic and nuclear physics, for instance to the position and velocity (more strictly, momentum) of a particle, but Bohr pointed out that it runs through the whole of our experience. For example, we can study how an organ of the body works in physical and chemical terms, but that means dissecting and analysing it, and we cannot at the same time study the organ as a whole, functioning in its normal way in the body.

Bohr was deeply convinced that complementarity expresses the true nature of the world, so Einstein *had* to be wrong. Einstein finally admitted that Bohr's position was logically tenable, but 'so very contrary to my scientific instincts that I cannot forgo the search for a more complete conception'.

Only a few physicists still have an interest in the controversy. For most of them it suffices that the algebra of quantum mechanics works. Yet its implications cannot be ignored when we ask deeper questions about the nature of the universe.

One further thing should be said. Quantum mechanics was developed to deal with the problems of the outer layers of the atom, but it is a general theory and should be valid for the nucleus as well. The application of the tunnel effect to alpha-radioactivity indicates that this is indeed the case. However, knowing the ground rules does not mean that we now have a full understanding of the nucleus, any more than knowing the laws of cricket means that we can predict the result of a test match, which depends on the players and the way they interact. Similarly, the behaviour of the nucleus depends on the properties of its constituent particles and the forces between them, as well as on the quantum-mechanical ground rules, and we are still rather in the dark about these forces. Nuclear science has therefore been built up much more through the discoveries of the experimentalists than through the predictions of the theorists, and the structure of the nucleus is still imperfectly understood.

2 The Golden Age

The year 1932 was a spectacular one in the history of nuclear science – *annus mirabilis* as a colleague of Rutherford's called it. Cockcroft and Walton's achievement of artificial transmutation was only one of a series of important advances. Even while these men were completing their 'atom smasher' at the Cavendish, James Chadwick made another far-reaching discovery in the same laboratory.

Two years earlier two German scientists, Walther Bothe and Herbert Becker, had performed alpha-bombardment experiments similar to Rutherford's, but with a target of the metal beryllium, and something new had appeared, different from the protons observed by Rutherford or anything else known at that time. People began to speak of a 'beryllium radiation', with a remarkable ability to penetrate through matter.

Then another unusual property of the 'beryllium radiation' was discovered by Marie Curie's son-in-law, Frédéric Joliot, and her daughter, Irène, at the Institut de Radium in Paris. At twenty-five, knowing little about radioactivity, Frédéric had become Marie's personal assistant, and a year later, in 1926, had married Irène. Not long afterwards an old friend said to him, 'You've come too late to radioactivity. The radioactive decay series . . . are known and there is hardly anything left to do . . .' How wrong can one be?

Undeterred, the Joliots set about exploiting the Institute's special asset, a substantial stock of radium. In 1929 they decided to prepare a large quantity of the highly radioactive element, polonium, from this in order 'to force important discoveries'. In due course they mixed the polonium with beryllium to produce a strong source of the 'beryllium radiation', and used it to test the effect of the radiation on atoms of the element hydrogen, chemically combined in the form of paraffin wax. They found that the radiation had a novel ability to knock individual hydrogen nuclei (protons) out of the paraffin wax with considerable force, setting them travelling at high speed.

But what was this mysterious 'beryllium radiation' with these unusual properties? Chadwick, spurred on by Rutherford, set about solving the riddle. He showed that the Joliots' phenomenon can be compared to a moving billiard ball striking a stationary one, and by calculating back he could work out that the 'beryllium radiation' consists of 'particles of mass nearly equal to that of the proton and with no [electric] charge'. He named the new particles 'neutrons', because they are electrically neutral.

The theoreticians heaved a sigh of relief at the new discovery because it gave them at last a reasonable starting-point for explaining the nucleus: protons and neutrons held together by very powerful forces. Rutherford had had an intuition along these lines as early as 1920, saying that particles like neutrons seemed 'almost necessary to explain the building of the nuclei of the heavy elements'.

Although it could not be appreciated at the time, the discovery of the neutron was also an essential step on the road to exploiting nuclear energy.

Later in the same year, 1932, Harold C. Urey and two colleagues in the US announced the existence of an isotope of hydrogen with atoms about twice as heavy as those of ordinary hydrogen. The presence of heavy hydrogen in ordinary hydrogen, and of heavy water in that most familiar of substances, ordinary water, was a considerable surprise to the scientific world. The proportion of heavy hydrogen (also called 'deuterium') is small in nature – only about one part in 5,000 – which largely explains why it went undiscovered for so long.

Still in 1932 came another American discovery. Four years earlier a gifted young Cambridge mathematician, Paul Dirac, had predicted on purely theoretical grounds that there should be a particle like the electron, but of opposite electric charge – positive instead of negative. Now Carl D. Anderson in the US vindicated this theory by finding Dirac's particles among the products of the cosmic rays that reach the Earth continually from outer space. Their existence was confirmed soon afterwards at the Cavendish, with improved technique, by Patrick Blackett (distinguished in the Second World War for applying science to naval and military operations) and Guiseppe Occhialini. They are known as 'positrons'.

The discoveries of 1932 gave two more particles that could be used in bombardment experiments, neutrons and deuterons, the latter being the nuclei of heavy hydrogen. Moreover a new kind of particle accelerator, the cyclotron, had been invented in 1930 by an

American, Ernest O. Lawrence, whose skill in getting machines of this kind to work was later to prove important in the race for the atom bomb. By 1933 the cyclotron was being used successfully to accelerate protons and deuterons to high energies, and indirectly to release neutrons, for bombardment experiments.

Table 2. Elementary particles known by 1932.

Particle		Mass	Electric charge
Light	Electron	m	$-e$
	Positron	m	$+e$
Heavy	Proton (hydrogen nucleus)	1836 m	$+e$
	Neutron	1839 m	zero

A proton and an electron, having equal but opposite charges, combine to give a neutral hydrogen atom, in which the nucleus, the proton, carries nearly all the mass. Protons and neutrons are the building bricks of atomic nuclei.

Two further particles were used by physicists at that time: deuterons (heavy hydrogen nuclei); alpha-particles (helium nuclei, emitted by some radioactive materials).

Cockcroft said of this amazing period, 'We were living in the Golden Age of physics, so rapidly did discoveries come along', and Chadwick expressed some of the spirit of it when he described their research as 'a kind of sport. It was contending with nature.'

Surprisingly few people were involved throughout the world: about a hundred in all in the two leading centres, the Cavendish and the Institut de Radium, perhaps twice as many in smaller groups in other laboratories. Research news flowed freely among them, and they had a sense of fellowship across national boundaries. People sometimes spoke of a scientific International.

Soon, however, Hitler's accession to power in Germany in 1933 cast grim political shadows across the scene. An early shock was the expulsion of Einstein from the Prussian Academy of Sciences because he was a Jew. All over Germany dismissals of Jews from their posts began, and many of them left the country, including scientists who would have been invaluable to the Nazis in the coming war.

Meanwhile advances in nuclear science continued. The year 1934 saw another tremendously important discovery, yet again in an alpha-bombardment. The first step came in the course of a series of experiments which Frédéric and Irène Joliot made with their powerful polonium sources. They noticed that, besides the now familiar protons and neutrons, some targets yielded positrons, the particles discovered by Anderson two years earlier in cosmic rays. This was the first time that positrons had turned up in nuclear reactions in the laboratory. When they reported these unexpected findings at a scientific conference in Brussels in 1933, they met with scepticism. They felt rather depressed, but no less a person than Bohr took the couple aside and encouraged them to stick at it. A few weeks later they achieved a major breakthrough.

'I irradiate this target with alpha-particles from my source,' Joliot told a colleague. 'You can hear the Geiger counter crackling. I remove the source: the crackling ought to stop, but in fact it continues.' What was happening was that the aluminium target was continuing to send out positrons. This lasted for a few minutes, the effect dying down meanwhile and finally disappearing.

This meant that the aluminium had become radioactive. It was the first example of artificial radioactivity, 'controlled alchemy' as a fellow-worker called it. A short-lived radioactive species (actually an isotope of phosphorus) had been produced by the action of alpha-particles on aluminium. Two other elements, boron and magnesium, gave similar results. The phenomenon of radioactivity, which had previously been virtually confined to a few rather exotic elements, had now been extended to some of the most ordinary and familiar elements known to the chemist.

Marie Curie was thrilled. Joliot wrote later: 'I will never forget the expression of intense joy which came over her . . . This was doubtless the last great satisfaction of her life.' A few months afterwards she died of leukaemia.

To one of his assistants, a young German by the name of Wolfgang Gentner, Joliot said, 'With the neutron we were too late. With the positron we were too late. Now we are in time!' A year later the Joliots were awarded the Nobel prize for their discovery.

Blackett called it 'an oddity of scientific history' that nobody previously had done what the Joliots did, either deliberately or by accident. The extra twist was, after all, merely to go on observing the target after removing the alpha-source. But once the ice was broken, further examples of artificial radioactivity were found in plenty. Cockcroft and his colleagues used their proton accelerator to produce it, and the Americans their cyclotron; the latter became for some years much the most powerful instrument available for the purpose.

Among those stimulated by the Joliots' results was an Italian,

Enrico Fermi, the man who, a bare eight years later, was to build the world's first artificial nuclear reactor. The thought struck him that neutrons might be better than alpha-particles for bombardment experiments. Having no electric charge they are neither repelled nor attracted by nuclei. Alpha-particles, on the other hand, are repelled by nuclei, because both have positive charges, and like charges repel. Alpha-particles therefore require a high energy to be effective, so as to overcome the repulsion and penetrate the nucleus. They can indeed only be used successfully when the nuclear charge and hence the repulsions are small, and this limits their effectiveness to just a few elements.

Fermi had just completed an arduous theoretical investigation, and was glad to turn for a time from abstruse mathematics to laboratory work. Neither he nor anybody else in Rome had experience of the techniques needed, but he boldly buckled to, and made his own Geiger counters (you could not buy them in those days) and prepared neutron sources with the aid of a gram of radium in the basement of the Public Health Office. Then he bombarded one element after another, systematically starting with the lightest, hydrogen, and working up. Nothing happened in the first six experiments, and Fermi nearly gave up, but the element fluorine in the seventh experiment produced a strong effect, as did many other elements after that.

Fermi called in several colleagues to help. He sent Emilio Segrè out with a shopping list to buy as many of the chemical elements as Rome could provide. Segrè was the first person ever to ask the city's chief chemical supplier for the rare metals, rubidium and caesium.

In all, Fermi was able to test over sixty of the ninety known elements, almost on a production line basis, and over forty of them became radioactive under the influence of neutrons. The Italians' first report on the subject was despatched to a journal in May 1934, only four months after the Joliots' pioneering work. This would have been a remarkably short period even for an experienced, well-equipped group, let along one starting from scratch. Yet it also emphasizes how essentially simple were the pre-war techniques.

Later that year the Rome group made another important discovery. They had been joined by Bruno Pontecorvo, an ebullient young man, newly graduated, who years later was to defect from Harwell to the USSR. Pontecorvo and Edoardo Amaldi, another of the team, were activating a silver tube by putting a neutron source inside it, and they got some odd results. For instance, more radioactivity was produced if the activation was carried out on a wooden table than on a metal plate. After several such experiments, partly at random, Fermi suggested carrying out the activation in a hole in a large block of paraffin wax. The activity now increased fantastically, a hundredfold, as if by some species of black magic.

Fermi was as surprised as the rest, but during the lunch break he worked out a possible explanation. His idea was, first, that the neutrons collide repeatedly with the hydrogen atoms in paraffin wax, and that this slows them down, and secondly, that slow neutrons produce much more activity than do fast neutrons. Hydrogen atoms, he reasoned, should be more effective than any others in decelerating neutrons, because they are almost equal in mass to the neutron. (This is not self-evident, but can be checked by calculations on billiard-ball type collisions.)

A simple test suggested itself: repeat the experiment in water, which contains about as many hydrogen atoms to the litre as paraffin wax. That very afternoon the neutron source and the silver tube were lowered into a goldfish pool in a garden behind the laboratory, and the same intense activity was produced as before. A further series of experiments showed that the effect was not confined to silver; most of the neutron-produced activities were enhanced by hydrogenous materials.

Materials such as water and paraffin wax which slow down the neutrons are now called 'moderators' (they 'moderate' the neutron velocities), and they are of great importance in nuclear reactors. This, however, lay in the future in 1934. The immediate significance of the Roman discoveries resided mainly in the fact that artificial radio-elements could now be made very simply in quantities that lent themselves to study or exploitation. It was not necessary to own a cyclotron. The chemists and biologists, as well as the physicists, began to find these materials very useful.

The following year the pace of discovery in Rome diminished, and Segrè asked Fermi why. It was partly that in their particular line of research they had skimmed the cream. Fermi, however, told Segrè to look at what was going on in the world, at Mussolini's ill-fated Ethiopian adventure, to say nothing of the activities of the Nazis in Germany. The fact was that they were all worried and were no longer devoting their minds wholeheartedly to science. Three years later, after winning the Nobel prize in 1938, the pressures drove Fermi to leave Italy for the US, because his wife was Jewish.

Groups elsewhere, however, enthusiastically followed up the

Italian discoveries, among them the men around Bohr in Copenhagen. By this time Bohr was a legend in his own country. As the foremost Danish savant, he lived in the Residence of Honour in the grounds of the Carlsberg brewery, with pilsner and lager on tap, and even the tram conductors knew about him. His colleagues would stand up respectfully when he came to the lunch room, while he stood shyly on the threshold. But none of this affected his personal friendships or his total absorption in pursuing, almost reverentially, the understanding of Nature.

When *La Ricerca Scientifica* containing Fermi's and his co-workers' papers arrived at Bohr's Institute, everyone gathered in some excitement round Otto Frisch, a young Austrian Jew, who was the only one able to read Italian. Frisch, a refugee from Nazism, had newly arrived at the Institute, at Bohr's invitation, after a year in London with Blackett, who had taught him nuclear techniques. He was destined to play a key part in the discovery of fission and in stimulating the atom bomb project.

The immediate reaction in Copenhagen to the Italian reports was, 'We need a strong source of neutrons ourselves.' Accordingly a public appeal was launched for a hundred thousand kroner (then about £5,000) to buy six-tenths of a gram of radium for Bohr's fiftieth birthday on 7 October 1935. This was mixed with finely ground beryllium to make the neutron source. Frisch was in charge of this particular job and soon he was using the source himself to study the passage of neutrons through different kinds of matter.

Bohr followed the results with great interest. Gradually a pattern emerged, but for some months it eluded explanation. Then towards the end of 1935 there was a colloquium at the Institute and Bohr, his mind active but puzzled, kept interrupting the speaker. Suddenly he stopped in the middle of a sentence and sat down as if unwell. A moment later he got up again smiling and said, 'Now I understand it.'

That was the beginning of a new picture of the nucleus, which Bohr and his colleagues elaborated over the next few years. He pointed out that the nucleus is a cluster of small spheres – the protons and neutrons – which tend to stick together when they touch, though not so firmly as to stop them moving around. But that is just what a drop of liquid is like – a lot of slightly sticky little objects (atoms or molecules) continually on the move. So one would expect the nucleus to have some of the properties of liquid drops.

The liquid-drop analogy can indeed be pushed to surprising

Table 3. Important dates in the 'golden age' of nuclear science.

1930	Lawrence invents the cyclotron.
1932	Cockcroft and Walton achieve a nuclear transmutation with the aid of a particle accelerator ('atom smasher'). Chadwick discovers the neutron. Anderson discovers the positron. Urey discovers heavy hydrogen.
1934	The Joliot-Curies discover artificial radioactivity.
1935	Fermi introduces the idea of moderators to slow neutrons down.
1936	Bohr propounds the liquid-drop model of the nucleus.

lengths. We can speak of particles from outside 'condensing' on it, or its own particles 'evaporating' from it. We can speak of raising its 'temperature' by increasing its energy content, which gives its particles a greater tendency to 'evaporate' from it. We can speak of a nuclear 'surface tension', which we can regard as helping to hold the nucleus together.

A few years later the 'liquid-drop model' was to provide a ready-made picture of the process of nuclear fission.

3 Fission

The radiochemists, particularly those in Berlin and Paris, took special note of the results reported from Rome on the bombardment of uranium with neutrons. These were more complicated than with any other element. Four if not five radioactive products had been recognized, and from their chemical behaviour the Italians thought that at least two belonged to hitherto unknown elements beyond uranium, the so-called transuranium elements. In this they were wrong. Though they did not know it they were observing the process of nuclear fission; the new species were fission products, which are isotopes of elements altogether different from the transuranium elements.

News of the peculiar new radioactive products reached two of the world's most experienced radiochemists, Otto Hahn and Lise Meitner, who was Frisch's aunt, on their return to the Kaiser Wilhelm Institute in Berlin from an international conference. Meitner had worked with Hahn ever since coming to Berlin from Vienna in 1907 for a two-year stay.

One of their assistants wondered how they could sleep a wink before checking the Italians' experiments, and an old colleague of Hahn's, Aristide von Grosse, increased the tension by writing from America expressing doubts about Fermi's transuranium interpretation of his results. 'We felt ourselves bound to find out which of the two was right, Fermi or Grosse,' said Hahn. They put all their other research aside to meet the challenge, and another radiochemist, Fritz Strassmann, joined them.

They soon discovered fresh complications. By 1937 they had a list of nine different species formed from uranium. One of these was identified (correctly) as a uranium isotope. That was quite normal, but the behaviour of the other eight seemed to support Fermi's transuranium hypothesis, though this raised serious nuclear physical difficulties.

Meanwhile in Paris, Irène Joliot-Curie was also attacking the problem, along with a Jugoslav physicist, Pavle Savitch. They discovered yet another species and studied it particularly carefully, with a view to identifying it beyond doubt. Curiously, it behaved like the element lanthanum, one of the 'rare earths', whose atoms are rather over half the size of those of uranium. We now know that it really was lanthanum, and that Joliot-Curie and Savitch were within a hair's breadth of discovering fission. But one unlucky experiment put them off the scent, and they mistook the new species for the very similar element, actinium. This seemed indeed more plausible because the jump from uranium to actinium is much smaller than that to lanthanum (a loss of 3 as against 35 positive charges).

Perhaps if Irène's husband, Frédéric Joliot, with his physical insight, had bent his mind to the problem, they would together have reached the right answer. He, however, was busy in many other directions, establishing a new base in the Collège de France, building nuclear accelerators, wrestling with the Government for finance. On top of that he was becoming increasingly active politically in left-wing organizations, partly to counteract Fascism and Hitlerism.

Of course he still kept in touch with the latest advances in nuclear science, and soon after the work just described he was at a scientific conference in Rome talking to Hahn. Hahn said that despite his great respect for Irene, he was going to repeat her experiments, and expected to show that she had made a mistake. When he and his colleagues did this, however, they thought they had confirmed the French claim, and extended it to two more 'actinium' isotopes and three 'radium' isotopes which were the precursors or parents of the three 'actinium' isotopes. The 'radium' behaved chemically like barium, and we now know that it *was* barium, but it is easy to confuse the two elements, and Hahn was expecting radium.

About this time the team in Berlin was disrupted by Germany's annexation of Austria, the Anschluss. Meitner was no longer protected by her Austrian nationality, but automatically became German, subject to the Nazi race laws. She was refused permission to leave the country, but with the help of Dutch friends she crossed into Holland illegally, without a visa, while Hahn waited tensely for news of her safe arrival. She was lucky; many others trying to leave were arrested. From Holland she went to Stockholm where she had been invited to settle, but she was far from happy in exile with neither the apparatus she wanted nor colleagues with whom she could discuss nuclear physics. Hahn missed her sadly, but continued with the research they had been doing together.

Lecturing on their uranium investigations in Copenhagen a short time later, Hahn was challenged by Bohr. As a nuclear physicist, Bohr could not see how uranium *plus* slow neutrons could possibly produce radium. Hahn replied that as a chemist he could not see how his material could be anything but radium. Bohr then suggested that perhaps they were dealing with a peculiar new transuranium element. Neither voiced the correct explanation, that the material not only behaved like barium, but *was* barium. 'This just shows how wildly impossible it seemed to regard barium as the product of the reaction,' commented Hahn later.

Nevertheless Hahn and Strassmann proceeded to the final check: they tried to separate their 'radium' from barium. The techniques were thoroughly familiar to them, but tedious to carry out, which was presumably why they had not done the experiments sooner. To their consternation the separation failed. As a further check they added a known radium isotope, formerly used in luminous watches under the name of mesothorium I, and tried the separation again. The genuine radium isotope behaved normally, but the species they were trying to identify obstinately remained with the barium. That was on Saturday, 17 December 1938, and Hahn wrote in his notebook, 'Exciting fractionation of radium/barium/ mesothorium.'

On Monday, 19 December, they started a confirmatory experiment. If their unknown really was barium and not radium, then its daughter should be lanthanum and not actinium, and this could be tested by a parallel separation to the one they had just carried out.

While this was in progress, Hahn wrote a long letter to Meitner, which is quoted in his book *My Life* (Macdonald, London, 1970). In it he said:

> It is now just eleven o'clock at night. At a quarter to twelve Strassmann will be returning so that I can see about going home. The fact is, there is something so odd about the 'radium isotopes' that for the time being we are telling only you about it . . . Our 'radium' isotope is behaving like barium . . . Perhaps you can suggest some kind of fantastic explanation.
>
> We all know that it [the uranium nucleus] can't *really* burst into pieces to form barium. But now we are going to see whether the 'actinium' isotopes formed by our 'radium' are going in fact to behave like actinium – or like lanthanum. All highly tricky experiments! But we must get at the truth . . .
>
> I have got to get back to the counters now.

Tuesday was the Kaiser Wilhelm Institute's Christmas party, but by the end of Wednesday the confirmatory experiment was finished. The 'actinium' was indeed lanthanum.

On Thursday, 22 December, Hahn and Strassmann wrote a short paper for the scientific journal *Naturwissenschaften*, describing their 'horrifying conclusion', as Hahn had called it in his letter to Meitner, a conclusion 'contradicting all previous experience' in nuclear physics. The editor, Paul Rosbaud, was so impressed that he made room for the paper in the next issue of the journal, even though other material was already set in type. The journal appeared on 6 January 1939.

Meanwhile, Meitner had received Hahn's letter. She was at Kungälv near Gothenburg, spending Christmas with Swedish friends. Her first reaction to Hahn's news was cautious, but she kept an open mind. 'We have experienced so many surprises in nuclear physics that one cannot dismiss this by saying simply, "It's not possible!"'

Her nephew, Frisch, came up from Copenhagen to join her for the holiday. He found her puzzling over the letter when he met her after his first night in Kungälv. He wanted to discuss a new experiment he was planning, involving a large magnet, but his aunt insisted on his reading the letter. He said later: 'Its contents were so startling that I was at first inclined to be sceptical . . . The suggestion that they might after all have made a mistake was waved aside by Lise Meitner; Hahn was too good a chemist for that, she assured me.'

Meitner and Frisch discussed the problem during a walk through the woods in the snow. The nucleus of the barium atom is not much more than half the size of the uranium nucleus; how on earth could the one be formed from the other? In all the nuclear processes known at the time, only small fragments were ever chipped off the nuclei. It would take a lot of small chippings to reduce uranium to barium, and there was not enough energy available for that. Nor could the uranium nucleus have been cracked in two; nuclei are not brittle like glass. As Bohr had suggested a few years before, they are more like drops of liquid, and it was this that gave the clue.

> Perhaps a drop could divide into two smaller drops in a more gradual manner, by first becoming elongated, then constricted, and finally being torn rather than broken in two? We knew that there were strong forces that would resist such a process, just as

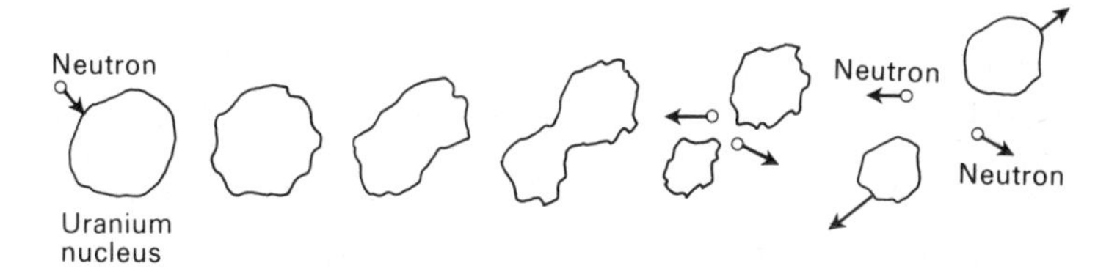

Fig. 5. A pictorial representation of nuclear fission.
The nucleus absorbs a neutron, becomes unstable (wobbly), develops a waist, and divides into two, at the same time releasing two or three secondary neutrons.

> the surface tension of an ordinary liquid drop resists its division into two smaller ones. But nuclei differed from ordinary drops in one important way: they were electrically charged, and this was known to diminish the effect of surface tension.
>
> At that point we both sat down on a tree trunk . . . and started to calculate on scraps of paper. The charge of a uranium nucleus, we found, was indeed large enough to destroy the effect of surface tension almost completely; so the uranium nucleus might indeed be a very wobbly, unstable drop, ready to divide itself at the slightest provocation (such as the impact of a neutron).

Pursuing this line of thought, they saw a possible snag. The two smaller drops into which the uranium nucleus divided would share the original electric charge on the nucleus, and – since like charges repel – the two parts would fly apart with great energy. The energy was easily calculated, and it was much larger than any encountered so far in nuclear physics laboratories. (It was about 200 million electron-volts.) Where could it have come from? The answer was that mass had been converted into energy in accordance with Einstein's $E = mc^2$ relation. The two smaller nuclei together weigh slightly less than the uranium nucleus from which they are formed. Meitner calculated the difference to be about one-fifth of the mass of a proton, and when this was inserted into Einstein's relation, the corresponding energy came to just the right value. So everything fitted! The uranium nucleus *did* burst into pieces Fig. 5).

After Christmas, Meitner returned to Stockholm, while Frisch travelled back to Copenhagen 'in considerable excitement' to report their speculations to Bohr. Bohr knew nothing so far, since *Naturwissenschaften* with Hahn and Strassmann's paper had not yet appeared.

> When I reached Bohr he had only a few minutes left [before leaving for the US]; but I had hardly begun to tell him, when he struck his forehead with his hand and exclaimed: 'Oh what idiots we all have been! Oh but this is wonderful! This is just as it must be! Have you and Lise Meitner written a paper about it?' I said, we hadn't yet but would at once, and Bohr promised not to talk about it before the paper was out. Then he was off to catch his boat.

The paper was drafted over the long-distance telephone and dispatched to *Nature* in London on 16 January 1939, with the title 'A New Type of Nuclear Reaction'. From the analogy with cell division in biology, Meitner and Frisch named the new process nuclear 'fission'. Accompanying their paper was a second note containing the results of a confirmatory experiment, suggested by a Copenhagen colleague, George Plaçzek, another Jewish refugee, in which Frisch demonstrated the very high energy of the two fragments produced by fission. Frisch called this a 'very easy' experiment; it took him only two days to put together the apparatus for it.

The two papers appeared on 11 and 18 February respectively. It was as well for Meitner and Frisch that they had acted quickly because others who saw Hahn and Strassmann's paper in *Naturwissenschaften* soon began to draw similar conclusions.

Bohr arrived in New York with his colleague Leon Rosenfeld on the very day that Meitner and Frisch posted their letters to *Nature*. On the boat they had discussed nuclear fission from every possible angle, but unfortunately Bohr had forgotten to warn Rosenfeld to keep the secret until the news was published. Bohr stayed in New York to see Fermi at Columbia University, while Rosenfeld went on ahead to Princeton, and there he let the cat out of the bag. (*Naturwissenschaften* had presumably not yet arrived in the US.) To Rosenfeld's dismay this unleashed a race among American physicists, mostly bent on proving the high energy of the fission fragments, but unaware that Frisch had already done just that. The

Physics Department at Princeton was said to be 'like a stirred up anthill'.

Everything came to a head at a conference on theoretical physics in Washington towards the end of January. Bohr had perforce to tell the whole story, starting with Hahn and Strassmann's discoveries; this he did on 26 January. It is related that some of those present dashed to their laboratories in full evening dress, even before Bohr had finished speaking, to get on the band wagon. Another tale is of a physicist watching his apparatus for evidence of fission fragments and simultaneously reporting over the telephone to a newspaper man: 'There's another one.' Seldom if ever has the scientific world seen another such scramble to be first with new discoveries. Bohr and Rosenfeld had some trouble in establishing the true priority in face of erroneous newspaper reports.

The effect of Hahn and Strassmann, Meitner and Frisch's work was like switching on a light in a dark room. Those who had been groping could now see clearly, and others rushed in to join them. New results started pouring in from Copenhagen, Cambridge, Paris, Berlin, New York, Berkeley (near San Francisco) – virtually every nuclear physics centre in the world.

Some looked back regretfully at what they had missed. In Cambridge, large electrical impulses caused by fission fragments had actually been seen in their apparatus, but dismissed as due to an electrical fault. Irène Joliot-Curie, who had so nearly anticipated Hahn and Strassmann's discovery, said, as Bohr had done, 'What fools we have been!'

Rather than bewail the lost opportunity, her husband, Frédéric Joliot, made sure that the Paris team would play a major part in the next act of the drama. When *Naturwissenschaften* with Hahn and Strassmann's paper reached him on 16 January, he spent a few days in intense thought. Like Meitner and Frisch, but independently of them, he decided that fission must be the explanation of Hahn and Strassmann's results, and he too realized that the fission fragments must have a very high energy. An experiment to demonstrate the latter was made in Paris on 26 January. This so convinced Joliot and his colleagues of the reality of fission that they dropped everything in order to follow up the consequences of the phenomenon.

4 Critical Experiments

Within days of the discovery of fission it occurred to a number of scientists that neutrons might be set free in the process. This idea led on to another, that here perhaps was the germ of a method for large-scale release of the vast energy of the atomic nucleus. There was talk of a 'super-bomb'.

The point is that if neutrons both initiate fission and are produced by it, there can be a chain of fissions. The secondary neutrons formed in fission go on to initiate more fissions; these liberate more neutrons; which cause yet further fissions; and so on and on (Fig. 6). The 'cascade messages' sometimes used by large

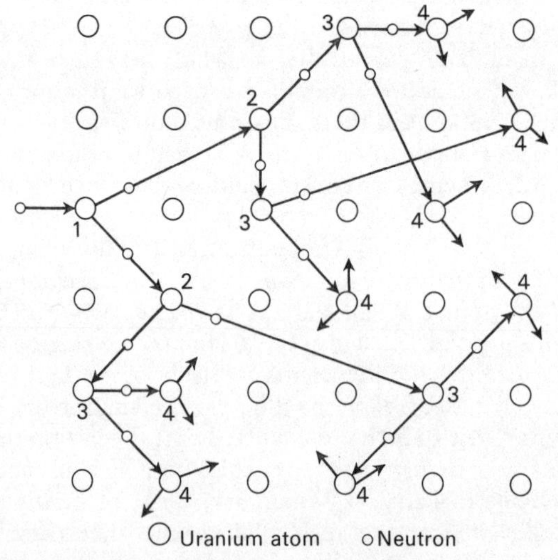

Fig. 6. Representation of a fission chain reaction.

The reaction is initiated by the neutron on the left, and the first four steps in the chain are indicated by numbers. Each fission is indicated as producing two neutrons, each of which causes a further fission. In reality many of the neutrons would be lost in non-fission processes and there would be very much less chain branching.

organizations in emergencies illustrate the point. One man rings up, say, five others to alert them, each of them alerts five more, each of them another five, and so it continues, with the numbers growing all the time, and rapidly becoming very large.

The idea of a chain reaction was already familiar to scientists in 1939 as the explanation of chemical explosions. If an analagous nuclear explosion were possible, they realised that it might be a million or more times as powerful.

This was a terrifying prospect, especially in a world heading rapidly towards war. As the French nuclear scientist Bertrand Goldschmidt has described in *L'Aventure Atomique*, the whole climate of nuclear research changed overnight:

> From one day to the next, atomic physics ceased to be the domain solely of fundamental research, the preserve of the isolated research worker. A new élite, that of nuclear scientists aware of their moral and political responsibilities, was about to appear on the scene, and play a crucial part in the lives of great nations.

Up to 1938, physics had been fun. Now the men in the 'ivory towers' suddenly found themselves custodians of knowledge that could change the course of history.

They had of course long been aware of the tremendous energy of the atomic nucleus, but had seen no means of tapping it. Rutherford, prescient though he usually was, had said publicly that 'anyone who looks for a source of power in the transformation of atoms is talking moonshine'. That was at the 1933 meeting of the British Association for the Advancement of Science. According to Heisenberg in his book *Der Teil und das Ganze* (Piper Verlag, München, 1969) neither he nor Bohr raised a dissenting voice when he expressed the same opinion privately in their presence. True, they were all thinking of experiments like those done by Cockcroft and Walton, where substantial amounts of electrical energy were expended in transmuting an infinitesimal quantity of matter, and to seek a net gain by such methods could fairly be called 'moonshine'. They did not foresee the achievement of a neutron chain reaction.

There was, however, one man who did, even at that early date, a Hungarian Jewish refugee from Nazism, Leo Szilard. He said that the idea flashed into his mind as the traffic lights changed from red to green, while he was walking along Southampton Row in London and pondering on Rutherford's 'moonshine'. He proceeded to work out the consequences in terms of nuclear power and bombs in remarkable detail, anticipating much that others were to re-invent several years later. It is all recorded in a British patent application dated 12 March 1934, which makes astonishing reading considering when it was written, but it had no immediate influence on events because Szilard, fearful of the possible consequences, kept it secret by assigning it to the Admiralty. From the notes he jotted down at the time, it seems quite likely that he might have forestalled Hahn and Strassmann in their discovery of fission if he had had the means to carry out his ideas for experiments; certainly he had noted the element uranium as a profitable case to investigate.

For some, the fears aroused in early 1939 were temporarily allayed by the belief that, even if a chain reaction could be achieved, it could not occur explosively. The argument for this came from no less a person than Bohr. It arose from a discussion at Princeton University, shortly after the discovery of fission, when Bohr had seen cogent reasons for its being the rare isotope of uranium, ^{235}U, rather than the predominant ^{238}U isotope, whose fission Hahn and Strassmann had observed. Plaçzek and an American, John A. Wheeler, who were taking part in the discussion, laid a bet, 1,846 cents to 1 cent, on Bohr's being right. (1,846 was the then accepted value of the proton : electron mass ratio.) Proof did not come until over a year later, in March 1940, when minute samples of partially separated ^{235}U and ^{238}U became available for testing. Bohr's speculation was confirmed, and Plaçzek sent Wheeler a cheque for $0.01.

The implication is that ^{238}U is actually a hindrance to achieving a chain reaction, because it mops up many of the neutrons without their causing fission. Its effect can be counteracted by decelerating the neutrons, since this boosts the amount of fission in ^{235}U. But with slow neutrons the whole process is too slow for an explosion, as Bohr pointed out; they take too long to travel from one uranium atom to the next. There can only be a 'fizzle', enough to disperse the uranium, and so stop the reaction.

Bohr was indeed quite right in thinking that ordinary uranium, as it occurs in nature, can never make a bomb. Suppose, however, that the ^{238}U could be removed, leaving pure ^{235}U? Six years later the Americans were to produce a bomb by doing just that. Bohr was not oblivious of the possibility, but in 1939 it looked like crying for the moon. No element except hydrogen, which is a particularly easy case, had ever been separated in bulk into its isotopes; the difficulty and expense appeared prohibitive.

Whether Bohr was right or wrong, the need in 1939 was for hard experimental facts to show the possibility or otherwise of a nuclear chain reaction. The first in the field seems to have been Joliot in Paris. He had made a passing reference to the idea of such a chain reaction, though without elaborating on it, in his Nobel prize speech in 1935. Now, as soon as he had satisfied himself by his own experiments that nuclear fission is a reality, he started work with two postgraduates, both of foreign origin, partly Jewish, who had become naturalized Frenchmen.

One of the postgraduates, Lew Kowarski, said that 'to be the first to achieve the chain reaction is like achieving the philosopher's stone. It's far more than a Nobel prize', while the other, Hans von Halban, Jr., has described the team as 'absolutely bent on creating a nuclear chain reaction that could be used for nuclear power'.

Joliot had been building up a new laboratory at the Collège de France. Now he had the right apparatus and the right men, he was free to turn from administration to research, and he could go all out. In the coming months the group was often to work twelve to fourteen hours a day in the laboratory.

The essential first step was to confirm the surmise on which the whole idea of a chain reaction depended, namely that secondary neutrons are set free in the fission of uranium. If the answer was yes, then the next step would be to determine how many. A minimum of one neutron per fission is required on an average if one fission is to lead to another, that one to yet another, and so on without the chain ever coming to an end. In fact there must be considerably more than one per fission, because many of the neutrons are lost in non-fission processes, and there must be at least one left after all the losses have been allowed for. To revert to the cascade message analogy, each person receiving the message must make one phone call to alert another if the chain is to continue, and if a proportion of the calls go astray or fail to alert the recipient, it will have to be more than one.

The experiments were more difficult than might be imagined. The problem was to detect the occasional secondary neutron amid the flood of primary neutrons needed to start the process. Here Halban was able to use techniques he had learned from Frisch in Copenhagen, where he had spent a year. Basically the method depended on putting a neutron source in the middle of a tank full of liquid containing uranium, and measuring how many neutrons

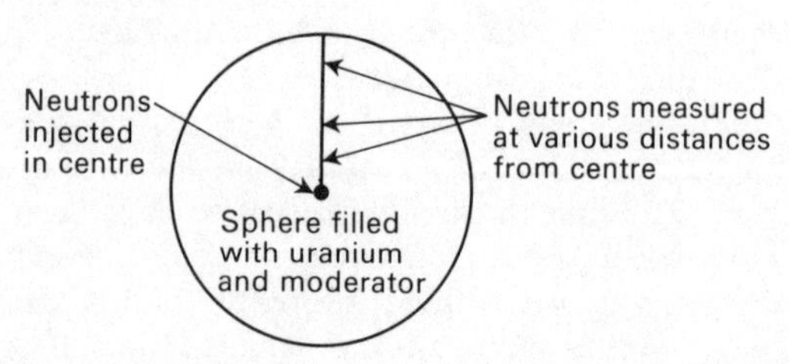

Fig. 7. Scheme for the detection of secondary neutrons from fission.

could be found in the liquid at different distances from the source (Fig. 7).

The number of neutrons falls off as the distance increases, just as a light gets fainter when you move away from it. The number drops still more rapidly if some of the neutrons are absorbed by the liquid, like light in a fog. From neutron intensity measurements it is therefore possible to determine the extent to which different liquids absorb neutrons (corresponding to the thickness of the fog), and in Copenhagen this is what Frisch, Halban, and a Dane called Henrik Koch had done.

Now the group in Paris filled their tank with a uranium (uranyl nitrate) solution, and then, for comparison, with an ammonium nitrate solution. The drop in the neutron intensity due to absorption was apparent in both cases. What especially interested the experimenters was the fact that out towards the edges of the tank (say twenty centimetres from the neutron source) there were considerably more neutrons in the uranyl nitrate than in the ammonium nitrate case. The excess must be the secondary neutrons they were looking for.

The team made a confirmatory experiment and then rushed into print. On 8 March, Kowarski went to Le Bourget airport to post a letter to *Nature*, which promised quicker publication than the French *Comptes Rendus*, and it appeared on 15 March.

The same discovery was made nearly simultaneously at Columbia University in New York by Fermi, Szilard, and their colleagues, but in view of the potential military significance, publication was temporarily withheld on Szilard's initiative.

Szilard had been on the point of withdrawing his secret British Admiralty patent when the news of nuclear fission broke, reviving all his old anxieties. One of his first thoughts was to keep knowledge of a possible chain reaction from reaching Germany, and he proposed a voluntary censorship of information on fission, so as to deny the Nazis the fruits of research in the countries they threatened, fruits they could pluck so long as the usual scientific practice of open publication continued. Bohr supported the idea, and Blackett assured Szilard that the Royal Society in London could be relied on to co-operate.

Szilard wrote to Joliot on 2 February to enlist him in the scheme, but to no avail. The French reaction was first one of surprise and then of rejection. Goldschmidt, who was working in the Collège de France at the time, has indicated their reasons:

> The free exchange of knowledge [in nuclear physics] had always been complete and sometimes even had the character of a race, where a few days more or less . . . could mean the difference between the glory of discovery and the lesser satisfaction of providing confirmatory evidence.

The Paris group were evidently in no mood to forgo their hope of glory, despite Joliot's concern about the advance of Nazism, and Szilard's letter failed to halt publication of their results. In extenuation it may be said that their minds seem to have been focused more on industrial than military applications.

Even after the 15 March letter from the French had appeared in *Nature*, Szilard still wanted to hold up the Columbia University results. These had already been dispatched to the *Physical Review*, but the editor had been requested not to publish them until the voluntary censorship issue had been resolved. In face of the French publication, Szilard was overruled by his colleagues and their letters appeared on 15 April. His plan had come to nothing. Over a hundred research papers on fission appeared in 1939, besides sensational newspaper articles. Later, when war came, thousands of scientists had of course to accept a total embargo on publication.

One of the *Physical Review* letters contained an estimate of two secondary neutrons per fission. So, one neutron to carry on the chain, and one to make up for losses; it might well be enough. A still more encouraging figure of 3.5 secondary neutrons (on average) was published by the French in a further letter to *Nature* on 22 April, though a recalculation some months later, in which a theoretical error was corrected, brought this down to 2.6; the figure accepted today is about 2.5. In Russia, too, there was a similar publication about this time.

Thus towards the end of April 1939, the first key steps had been taken towards a chain reaction in uranium, and they had been broadcast to the world.

In view of the serious implications the Columbia University scientists had informed the US Government as soon as they had obtained their first results. The move, suggested by another Hungarian exile, Eugene Wigner, took the form of a meeting on 17 March between Fermi and a group of naval and military scientists in Washington. This was the first approach to officialdom on the subject anywhere in the world, but very little came of it, despite an encouraging response, mainly because Fermi himself was not yet very convinced. Even after fission had been discovered he is reported to have said 'Nuts!' to the idea of an atom bomb and to have admitted only a 'remote possibility' of a chain reaction.

The French did not approach their Government at this stage, but took out patents on the applications of their discoveries. Their motive was not personal gain but patriotism; they transferred the ownership of the patents to public scientific bodies with the aim of ensuring that France would be in the lead if and when nuclear power was developed industrially.

Joliot also got in touch with Edgar Sengier, President of the Union Minière du Haut-Katanga, a Belgian company engaged in extracting radium from uranium ores from the Congo (now Zaïre). He obtained uranium for his experiments, and proposed a joint uranium bomb project in the Sahara.

Scientists in Britain and Germany were first alerted to the possibility of a chain reaction by Joliot, Halban, and Kowarski's letter of 22 April in *Nature* (which reached them sooner than the 15 April *Physical Review*). They got in touch with their Governments immediately.

In London, George Thomson of Imperial College, son of the famous J. J. Thomson, consulted his colleagues, and within four days several Government departments had been put in the picture. Uranium supplies were felt to be the most urgent aspect. The only major stocks anywhere in the world were believed to be those held by the Union Minière, and it was decided that an effort should be made to reserve them for Britain, and to keep them out of German hands. An approach was therefore made to Sengier, who was very co-operative. He already knew about the potentialities of uranium from Joliot, and promised to inform the British of any abnormal orders, but though he had the residues from radium extraction, he had little refined uranium on hand. The Dutch also got in on the act and procured eight tons of uranium oxide from Sengier; it remained hidden in a cellar in Delft throughout the German occupation, and was one of the resources available to start a Dutch–Norwegian nuclear project after the war.

Action was also taken in Britain to start a co-ordinated research programme in different universities, even though several leading scientists expressed scepticism about the whole idea. Sir Henry Tizard, adviser on air defence, spoke of odds of a hundred thousand to one against a successful military application, and Frederick Lindemann (later Lord Cherwell), perhaps influenced by Bohr's reasoning, told Churchill that exploitation would take some years, and might not after all give exceptionally powerful weapons.

German scientists were as quick off the mark as Thomson. When the crucial 22 April issue of *Nature* appeared, there were two independent approaches to the Government, one by the physicists Georg Joos and Wilhelm Hanle at Göttingen University, and the other by the physical chemists, Paul Harteck and Wilhelm Groth at Hamburg University. Harteck's interest in the subject stemmed from five years earlier, when he had worked under Rutherford at the Cavendish.

Joos and Hanle wrote to the Ministry of Education, where Professor Abraham Esau, a Nazi supporter despite his Hebrew names, convened a meeting almost immediately, on 29 April, attended by several eminent physicists. One of the principal results was the commandeering of the limited stocks of uranium in Germany. News of the meeting leaked out to Britain, causing some measure of alarm.

Harteck and Groth sent a letter to the War Office on 24 April, saying: 'We take the liberty of calling to your attention the newest development in nuclear physics, which in our opinion will probably make it possible to produce an explosive many orders of magnitude more powerful than the conventional ones.'

After touring the administrative machine, this missive ultimately reached Kurt Diebner, a nuclear physicist by training and like Esau ready to serve the Nazi regime. He set himself up as the head of a nuclear research office in the Army Ordnance Department, despite his boss's saying, 'Do give up your atomic poppycock.'

There were thus two rival initiatives in different ministries, each in the hands of an ambitious man who saw a nuclear programme as a means to his own advancement. During the summer of 1939 Diebner edged Esau out, giving himself a clear field to launch Germany's wartime project. Meanwhile, during this infighting, scarcely any scientific or technological work on the exploitation of nuclear fission seems to have been done in Germany.

Russian scientists were also following the news of fission in the scientific journals, but apparently in a much more detached spirit. They were academically interested in the physics of chain-reacting systems and discussed them as a means to industrial power production, but remarkably enough they seem to have ignored the military applications, even though they investigated explosive chain reactions as a theoretical problem. There was a committee in the Academy of Sciences to study the 'uranium problem', but no real attempt to involve the Government. Nor was there any censorship; newspaper articles on atomic energy appeared freely. This carefree state of affairs seems to have persisted until the German invasion in 1941 brought Russian nuclear research to an abrupt halt.

Yet another of the future belligerents was keeping a watching brief – Japan. She had at that time a few outstanding nuclear physicists. One of them, Yoshio Nishina, had spent some years with Bohr in Copenhagen. Another, Ryokichi Sagane, went to Berkeley to learn about cyclotrons from Lawrence. The Japanese were undoubtedly able to theorize about the possible applications of nuclear energy as intelligently as their opposite numbers in Europe and America.

On the experimental side, the next logical move after the measurement of the yield of secondary neutrons in fission seemed to be to build a structure – a 'pile' or 'reactor' – in which there is a self-propagating chain process. This became one of the first objectives in Britain, France, and the US, and later in Germany. Hope was pinned mainly on slow neutrons, which cause so much more fission than fast, so uranium was combined with water or paraffin wax to slow down the neutrons. It was not thought that this would lead directly to a bomb, but it seemed the natural development, and it might lead to nuclear power.

Was a nuclear reactor possible? Or would the chains in any possible structure merely peter out? We would then have had a situation similar to that of a species of animal whose numbers dwindle until it becomes extinct, whereas what we are looking for is a population explosion of neutrons.

In such cases everything depends on the number and fate of the offspring – in the nuclear case, the secondary neutrons – in successive generations. It is a matter of birth-rate versus mortality-rate. By April 1939, as already described, it was known with fair certainty that the neutron birth-rate was high enough, but the mortality-rate, the extent of the neutron losses, was in doubt.

The question is: does the number of neutrons increase or decrease from one generation to the next? It is convenient therefore to consider the ratio of the numbers in successive generations. This is called the neutron multiplication factor, and is usually denoted by k. If, for instance, the number doubles from one generation to the next, $k = 2$, and if it increases by 5 per cent, $k = 1.05$. Any increase, be it noted, is at compound interest, and if unchecked may lead to an explosion, whereas a decrease, with k less than one, implies that the chains will peter out.

The would-be nuclear reactor builders of 1939 therefore measured the k-values for the types of structures or assemblies they were investigating, seeking all the time to achieve a k-value greater than one. The experiments involved studying the fate of neutrons injected either into the assemblies themselves or into their different constituent materials.

There is then a further consideration, for which an ordinary bonfire provides a parallel. A bonfire must be built up to a certain size before it will 'go'; if it is too small, too much of its heat leaks away to the surroundings instead of sustaining the combustion. Similarly a nuclear reactor must be built up to a certain minimum size, known as the 'critical size', in order to avoid too much leakage of neutrons to the surroundings. Below the critical size it is quiescent; above it, it can so to speak blaze up. The critical size marks the point at which the assembly is exactly in balance, the total neutron losses (wasteful absorption *plus* leakage) being precisely equal to fresh production. The same concept applies to an atom bomb.

Thus there are two requirements to be met before a self-sustaining chain reaction can be achieved:

- The k-value must be greater than one.
- The critical size must be reached.

If the k-value is less than one, the chains will always peter out. No increase in the size of the assembly will ever suffice; there is no critical size. If the k-value is greater than one, however, there is always some point at which the assembly will 'go critical' or 'reach criticality' as it is built up to larger and larger sizes. (For the technically minded it should be made clear that k is used here to denote the neutron multiplication factor in an infinitely large assembly, when there are no leakage losses, i.e. it is in fact k_∞.)

Francis Perrin, who had joined Joliot's group, is sometimes credited with being the first to introduce the important idea of a critical size in the spring of 1939, but he was in fact five years behind Szilard, who had included it in his secret British patent in 1934. Perrin certainly conceived it independently and applied it to the kind of assemblies the French were studying. He estimated that some forty tons of uranium would be needed before these assemblies would reach criticality, so that they would be quite large structures. His figure, though based on skimpy information, was of the right order of magnitude; the first nuclear reactor three and a half years later contained about fifty tons of uranium.

The assemblies investigated in those early days were generally quite crude. Fermi and Szilard started by simply dissolving a uranium compound in water. The French filled copper spheres of increasing size with wet uranium oxide and lowered them into a water tank. Thomson and his colleague Philip Moon made similar experiments, and also tried spheres of uranium oxide alone, without any water, with the idea of achieving a fast neutron chain reaction. Uranium metal would have been preferable to uranium oxide, but at that time it was a chemical curiosity, existing only in very small quantities; Mark Oliphant at Birmingham University had begun to study how it might be made in bulk.

Each of the three groups in the field realized, independently, as their work proceeded, that instead of dispersing the uranium more or less uniformly in the moderator, it was better to separate the two materials. Fermi tried suspending about fifty cans of uranium oxide in a tank of water, while the French tested the reverse scheme of lumps of moderator (paraffin wax in this case) spaced

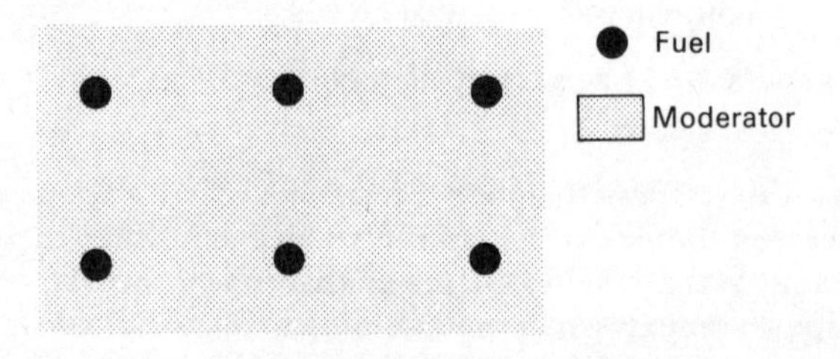

Fig. 8. Lattice arrangement of fuel (e.g. uranium) and moderator.

Originally the fuel was arranged in lumps in the moderator. Today the fuel is usually in the form of rods through the moderator. The general idea is that a neutron leaving one lump or rod of fuel should be fully slowed down before reaching the next.

out in uranium oxide. Such an arrangement is called a 'lattice' (Fig. 8). It was an important invention, used in nearly all nuclear reactors today, yet it is not known who thought of it first.

The main reason for having a lattice is that the neutrons are most vulnerable to wasteful capture by uranium itself when they have been partially but not completely slowed down. In the 1939 experiments it was therefore desirable to arrange that, so far as possible, a neutron produced in uranium travelled through a sufficient thickness of moderator to slow it down completely before it encountered any more uranium. For example, in Fermi's type of lattice, a neutron from one can of uranium oxide should preferably have been slowed down completely before it reached the next can.

It also dawned on the different groups that water or paraffin wax might be a bad choice of moderator, because the hydrogen atoms they contain, though ideal for slowing down the neutrons, also cause serious neutron losses by acting as neutron absorbers. Plaçzek pointed this out to Fermi and Szilard when he visited them in June 1939. Szilard thought of graphite as a possible alternative and proposed a crash programme, but Fermi had meanwhile gone away for the summer and was in any case more interested in working on cosmic rays. Szilard lacked the resources to go it alone, and there was in consequence an interregnum of about nine months when nobody in the US was trying to produce a chain reaction.

Plaçzek also put the point to the French, but for the time being both the British and the French continued working with uranium oxide/water or uranium oxide/paraffin wax assemblies. In August the French obtained a significant success with a fifty-centimetre sphere of wet uranium oxide in a tank of water. Following their usual practice they inserted a neutron source into the centre of the sphere, and measured the number of neutrons in the tank outside. From their results they were able to claim evidence of a chain reaction. The chains were very short, rapidly petered out, and were very far from producing useful amounts of energy, but they were chains nevertheless. This was the French group's last experiment before war was declared, and their report was seen in due course not only by the British and the Americans, but also by the Germans; thereafter they kept their work secret.

One other important paper on fission was published that August, by Bohr and Wheeler. In it the authors elaborated the ideas that had led Bohr to believe that it is the ^{235}U rather than the ^{238}U isotope that undergoes fission with slow neutrons. The theory in the paper enabled them to predict what other nuclear species might undergo fission with slow neutrons, including species not yet discovered. Among these was the principal plutonium isotope, ^{239}Pu. The paper therefore pointed the way to one of the major developments of the coming war, the plutonium bomb, and it was available to everyone in the open literature.

5 Germany's Early Wartime Lead

By the time Hitler started the Second World War by invading Poland in September 1939, much of the excitement about nuclear fission had died down.

In Britain, Thomson's efforts languished because, even if his experiments were successful, it was not obvious that they would lead to a weapon, and other defence tasks were urgent. Moreover, his results were discouraging: the highest k-value he was ultimately able to report was only 0.8, far below the key value of unity. A review by Chadwick at Liverpool University that autumn tended to confirm the gloomy view. He thought that a bomb, if possible at all, would require at least a ton of uranium – and how could so much material be brought together into one mass without blowing up prematurely? However he stressed that he was arguing from ignorance; there was too little basic scientific information.

In the US, Szilard had failed to keep the experimental programme at Columbia University alive. Frustrated, and motivated as ever by his fear of a Nazi atom bomb, he cast around in his mind for other ways of stirring up the Americans. The outcome was the celebrated letter from Einstein to President Roosevelt, warning of the potentialities and dangers of nuclear chain reactions, which Alexander Sachs, an economist with the ear of the White House, took to Roosevelt on 11 October 1939. Roosevelt said, 'Alex, what you are after is to see that the Nazis don't blow us up,' and then, 'This requires action.' He appointed an Advisory Committee on Uranium under Lyman J. Briggs, Director of the National Bureau of Standards. The Committee reported within a few days to the effect that nuclear power and nuclear explosions were possibilities, but still unproved. Thereafter, however, they became becalmed, and little more happened for some months; as yet few in America had a sense of urgency.

The French, like the British, had decided that a bomb was too remote a proposition, but they pursued their researches in hopes of developing nuclear power. They foresaw a possible application to submarines, since nuclear power requires no oxygen, enabling a submarine to remain submerged for indefinitely long periods, but their eyes were mainly on peaceful uses. To ensure continuance of their work in wartime, Joliot approached the French Government and won the enthusiastic support of Raoul Dautry, the Armaments Minister, who granted him 'exceptional facilities: unlimited credit and the possibility of recalling from the army any co-worker he may require'. Joliot was formally called up, bringing the programme under military auspices, but otherwise it was business as usual at the Collège de France until the German invasion the following May.

The group's primary aim throughout was to achieve a self-sustaining neutron chain reaction. In August they had succeeded in demonstrating short chains, and even though these petered out, the result gave them hope. Their optimism was somewhat dashed the following month when Kowarski corrected a theoretical error they had been making and showed that the k-value of their system was much less than unity. This, of course, was consistent with the chains failing to propagate, but it also showed that success could not be achieved simply by a bigger experiment of the same kind. A further trial on a larger scale only confirmed this.

One way out of the difficulty would have been along the lines indicated by Bohr's reasoning: to increase the proportion of the ^{235}U isotope in the uranium. The increase required was correctly estimated to be quite modest, only from 0.71 per cent to 0.85 per cent, and Joliot ordered some equipment to attempt an isotope separation. He did not pursue this line, however, because it looked a very arduous route to a chain reaction.

The other lines that were open were to try the lattice arrangement mentioned earlier, and to change the moderator. The French were able to show some limited benefit from using a lattice, but not enough for their purposes, so they turned to alternative moderators in place of water or paraffin wax.

Two main options presented themselves: heavy water and carbon in the form of graphite. Carbon in the form of solid carbon dioxide ('dry ice') was also tried in Paris and later in Hamburg. It is an ingenious idea for demonstrating a chain reaction, because dry ice is in effect a very pure form of carbon, whereas graphite may be contaminated with strong neutron absorbers that break any neutron chains. However dry ice would be useless in a reactor actually operating and generating an appreciable amount of heat since it would be lost by evaporation. There are a few other possibilities, such as the rare element beryllium, but none that could readily be used on a large scale in a reactor.

Table 4. Moderators.

Used in nuclear reactors	
Water	Cheap, but absorbs relatively large numbers of neutrons.
Heavy water	Ideal, apart from expense.
Graphite	Good, but special care is needed to avoid neutron-absorbing contaminants.
Suitable only for experimental purposes	
Paraffin wax	
Solid carbon dioxide ('dry ice')	

Of the different investigators only Fermi in America was to adopt graphite. The British dropped reactor-building altogether. The French mistakenly thought that graphite absorbed too many neutrons, and that they were forced to select heavy water, of which there appeared to be just enough in Norway for their initial experiments. They planned to test out heavy water in the spring of 1940. The Germans made the same error as the French and it conceivably lost them the chance to be first with a nuclear reactor, because they were always dogged by insufficient heavy water.

Up to this point the German project had been making excellent progress. On the outbreak of war, Erich Bagge, who worked under Heisenberg at Leipzig University, received his call-up papers and duly reported in Berlin, fearing that he was to be sent to the front. To his relief he was met by a fellow-scientist, Diebner of the Army Ordnance Office, and asked to help in setting up a secret conference on a prospective uranium project. This took place on 16 September, and was attended by many of Germany's foremost nuclear scientists, including Hahn, Bothe, Geiger, and Harteck, all of whom have been mentioned in earlier chapters. There was a further meeting a couple of weeks later, with the important addition of Heisenberg himself and one of his pupils, Carl Friedrich von Weizsacker. Weizsacker was a man of unusual motivation. He had read physics not so much for its own sake as to provide a basis for philosophical studies. He now wanted to join the 'Uranverein' (uranium club) because he saw nuclear science as a future political football, as well as a means of avoiding the call-up. He was to play a significant part in the relations between the 'Verein' and the Government.

By the time of the second of the two conferences, Heisenberg had already got certain essential points clear in his mind, in particular the distinction between the routes to nuclear weapons and to nuclear reactors. He foresaw that the former might be achieved by isolating the rare ^{235}U isotope and relying on fast neutron chains, and the latter by natural uranium/moderator mixtures and slow neutrons.

The conference drew up a national plan for the 'exploitation of nuclear fission', placed restrictions on publication, and established a centre for a 'Nuclear Physics Working Group' in the Kaiser Wilhelm Institute for Physics in Berlin, which was requisitioned for the purpose by the Army. The original proposal was to bring all the scientists in the project together in Berlin, but this was strenuously resisted, so that in the end there were some five participating and often competing groups in different cities. Nevertheless, the project got off to a very good start.

An incidental consequence of these decisions was the replacement of the head of the Kaiser Wilhelm Physics Institute, Paul Debye, because as a Dutchman he could not be put in charge of secret work under the Army's direction. Faced with the alternatives of naturalization or resignation, he chose to emigrate to the US, which he was free to do since the Netherlands was not yet at war. There he contributed to reawakening the Americans' interest by telling them of the existence of a large project connected with uranium in his former institute in Berlin.

Diebner, with the military behind him, saw his chance and moved into Debye's office. To counter him, the Institute's physicists arranged for Heisenberg to become a consultant and to pay regular visits from Leipzig. Though Diebner remained nominally in charge, Heisenberg gradually came to dominate the Uranverein by sheer scientific ability.

At the outset of the project, Heisenberg took on the task of writing a survey of the whole subject, which he submitted to the War Office on 6 December. His chief sources were American, British, and French scientific journals containing the very information Szilard had attempted to keep secret. To this he added his own original thinking, producing what was surely the best review of the state of the art then available anywhere. The ideas of a critical size, the neutron multiplication factor, heavy water and graphite as moderators, and the segregation of uranium from the moderator, e.g. in a lattice, were all included, besides the question of how to control a reactor to prevent the chain reaction running away, a topic which also preoccupied Halban in France.

Heisenberg's review showed the way forward: separation of the uranium isotopes with a view to a ^{235}U bomb, and attempts to achieve a slow neutron chain reaction with a view to a reactor, together with the nuclear measurements necessary for the latter objective.

Work on the first of these topics had already been started by Harteck. Initially he had in mind obtaining a few grams for laboratory experiments, but soon the Germans were thinking of much larger quantities. They were indeed the first to entertain seriously the idea of separating the uranium isotopes on a large scale, the idea that was to produce the Hiroshima bomb in 1945.

Harteck proposed to use a process which had recently been developed by two of his compatriots, Klaus Clusius and Gerhard Dickel, and applied with considerable success to neon, chlorine, and other gaseous elements. The element is introduced in gaseous form into a vertical tube heated along its central axis, for instance by a hot wire. The lighter isotope then tends to concentrate in the hot region; furthermore, because it is hot it rises up the tube. The heavier isotope meanwhile moves preferentially to the cool region by the wall of the tube, where it sinks. Thus the two isotopes tend to move in opposite directions, and a separation is achieved (Fig. 9). The technique is known as gaseous thermal diffusion.

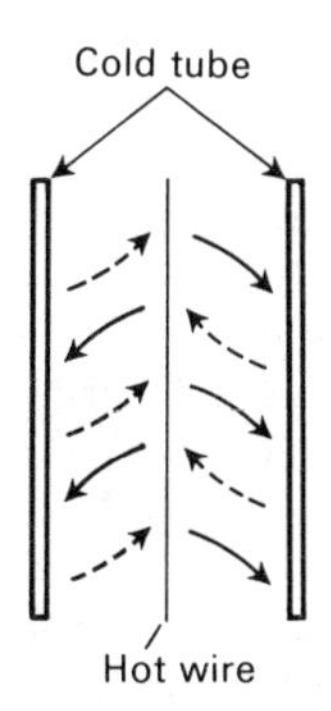

------→ Paths of lighter atoms – towards wire and upwards.

———→ Paths of heavier atoms – towards tube wall and downwards.

Fig. 9. The principle of thermal diffusion.

To apply it to uranium, Harteck required a gaseous uranium compound, and the only candidate appeared to be uranium hexafluoride. This is a nasty, corrosive material, and its molecules are not particularly suitable for the technique in question. Harteck and his colleagues plugged on with their experiments for some time but by early 1941 they had only obtained a slight separation on a small scale, and they gave up this particular line, mainly, it seems, because they were plagued by decomposition of the hexafluoride at the rather high temperatures needed at the centre of the separation tubes, though the process is in any case an inefficient one for uranium.

The main thrust of the German project was, however, towards demonstrating a nuclear chain reaction. Under Heisenberg's guidance the task was tackled systematically, starting with a foundation of theory and basic nuclear data. To some extent this was forced on the scientists by delays in obtaining supplies for large-scale experiments.

The materials they particularly needed were uranium and heavy water. Since the French had developed a similar interest, these substances became objects of war. Recognizing this, the French Government asked the Belgians early in 1940 to send their stocks of uranium to the US in case of a German invasion. The Union Miniere dispatched a large consignment, later used by the Americans, but still larger quantities remained in Belgium, and were in due course commandeered by the Germans. These were refined and chemically processed by Germany's excellent chemical industry, and in the middle of the war the German project was actually better off for uranium than the American.

At first, however, especially in the spring of 1940, the Germans were short of refined uranium. During this period, the indefatigable Harteck hit on the idea mentioned earlier, and already tried by the French, of using dry ice (solid carbon dioxide) as a moderator. He planned to see if a nuclear chain reaction could be observed in a uranium oxide/dry ice assembly. The work had to be done before the weather warmed up and the dry ice was needed for the refrigeration of food, so Harteck made an urgent plea to Diebner for uranium oxide. Heisenberg however wanted to reserve the limited stocks for his own experiments, and was unsympathetic to Harteck's 'suck it and see' approach. After tantalizing delays Harteck had to make do with 180 kilograms of uranium oxide. He embedded this in 15 tons of dry ice to form a lattice, and made a few measurements in early June, but was unable to detect any neutron multiplication.

He knew he had been baulked by too little uranium. Was it because the physicists did not accept a chemist as a full member of the Uranverein? Whatever the reason the consequences were far-reaching, because success by Harteck would have shown that pure carbon is a satisfactory moderator, and hence that a uranium/graphite reactor could succeed if the graphite was sufficiently pure.

The physicists meanwhile set about measuring the absorption of neutrons in potential moderator materials. Heisenberg's 1939 review had already virtually eliminated ordinary water and paraffin wax, so attention was turned to heavy water and graphite. The former was found to be excellent, absorbing scarcely any neutrons.

The graphite measurements were entrusted to Bothe, a most reliable experimentalist. His first result was discouraging, but he thought his graphite might have been insufficiently pure. However a later result with supposedly ultra-pure carbon was even worse. Such was Bothe's reputation that nobody questioned his work or sought to verify it; nor did the Uranverein know that Fermi in America had obtained a much more hopeful result. Yet a fateful decision depended on Bothe's work, for it led the Germans into abandoning graphite as a moderator, and so – unlike the Americans – missing the easiest route to a nuclear reactor. What went wrong with Bothe's work has never been clear; it can only be assumed that, despite his care, his material was impure. Harteck's experiments, if given full support, could have averted the calamity.

Having rejected graphite, the Germans pursued the idea of a uranium/heavy water reactor. For this they needed a substantial quantity of heavy water, and there was only one place in the world where it might be obtained, the Vemork plant belonging to Norsk Hydro near Rjukan in Norway, yielding about ten kilograms a month as a by-product of ammonia production for fertilizers. Norsk Hydro had links with the big German chemical concern, IG Farben, and IG representatives visited the Norwegians in January 1940 with a German Government order for their entire heavy water stocks plus a request to step up production tenfold. The Norwegians were astounded, but the Germans would not explain.

The French learned of the German interest in heavy water from an intercepted cable. There could only be one explanation: the Germans were trying to build a nuclear reactor! Since the French themselves had just decided to use heavy water in their own programme, they sent a representative to Norway immediately. The man chosen was Jacques Allier, who had a financial link with Norsk Hydro and was a secret service agent to boot. He told the Norwegians that heavy water was very important to the French war effort, and obtained the entire stock, 185 kilograms, free of charge, leaving none for the Germans. He also obtained first refusal of future production.

When the Germans got wind of the French coup they saw only one explanation: the French were trying to build a nuclear reactor! They tried to intercept the heavy water on its way to Paris, but they were foiled. They had been tricked into believing that it had been loaded on to an Amsterdam plane, which their fighters forced down in Hamburg, whereas it had actually been put aboard an Edinburgh flight standing alongside. The priceless, unique material finished its journey to Paris on 16 March.

The French intended to use the heavy water in experiments like those they had made with ordinary water as moderator, but events overtook them. The Germans broke through the French front on 16 May, and Dautry instructed Joliot to safeguard the heavy water. It was taken south to Clermont-Ferrand where Joliot planned to continue his experiments, but soon the Germans were threatening the whole country. On 16 June Allier warned Joliot that the situation was desperate. It was decided that Halban and Kowarski should take the heavy water to England, and they left from Bordeaux in a British ship, the *Broompark*, on 18 June. Joliot followed to Bordeaux, but missed his colleagues there, and decided that his duty was to stay in France.

Before leaving Paris, Joliot had burned all the papers relating to fission research, apart from a few essential ones he had taken to Clermont-Ferrand, but to no avail, because the Germans captured a set of his progress reports to the French Armaments Ministry.

This was but one of a series of nuclear assets that fell in quick succession into German hands as they conquered country after country in the first half of 1940. There were also the Copenhagen cyclotron, the Norsk Hydro heavy water plant, the large Belgian stockpiles of uranium, and the nearly finished Paris cyclotron. Only the existing stock of heavy water had eluded them, and with Norsk Hydro undamaged that seemed only a temporary set-back. By the middle of the year, Germany, her heavy industry intact, was very well placed to develop her nuclear project. The cyclotrons in particular filled a gap, and the French instrument was completed, but curiously enough no use was made of the Danish machine.

In terms of organization, resources, and scientific understanding the German project was the strongest in the world at that time, but it was weak in the motivation of its scientists. A German victory in the war meant a Nazi victory, and this was a prospect some of them regarded with dismay, while others were at best lukewarm. Hahn was an anti-Nazi, and quietly opted out of the war effort to do academic work on the products of fission. Gentner, derided on Gestapo files as having 'democratic ideals', was sent to Paris to run the Collège de France laboratory and get the cyclotron finished. From the Nazi point of view this was a mistake, because he knew and respected Joliot from the time he had worked with him in 1934. The two immediately entered into a private arrangement, restricting the laboratory so far as possible to fundamental and non-military research, and Gentner covered up for Joliot's activities in the Resistance.

Heisenberg, on whom the German project depended, had had a brush with the Nazis over the theory of relativity. The Nazis wanted to forbid the teaching of relativity on the grounds that its author, Einstein, was a Jew; but this was ridiculous, because modern physics is unteachable and indeed unthinkable without it. Heisenberg wrote an article in its defence in Hitler's newspaper *Das schwarze Korps*, and was denounced for his pains as a 'white Jew' by a strongly Nazi physicist, Johannes Stark. Himmler, who was a family friend, rallied to Heisenberg's support, but Heisenberg still came under fire from time to time. It is small wonder that he regarded Nazi race theory as dangerous idiocy.

This must have influenced his attitude to the Uranverein. Instead of seeing it as a contributor to the war effort, he and Weizsäcker used it to keep some of the best young physicists out of the armed forces, with an eye to the post-war period. To do this they had to engage in a delicate political game. They had to strike just the right balance between 'An atom bomb is possible' and 'It will take a long time'. The first statement served to ensure the continued existence of the project: the second, to hold off pressure for results. There was no actual dishonesty; the two statements summed up their own appraisal of the situation.

Even the faithful Nazi, Esau, advised soft-pedalling on the bomb for fear that Hitler might put them all behind barbed wire until he got it.

After the war the Uranverein made a virtue of the fact that they had directed their efforts to a nuclear reactor rather than nuclear explosives. It is true that, except at the outset, there is virtually no mention of atom bombs in their reports, and when asked in the middle of the war to give an account of their work to senior officials, they did not include a paper dealing with weapons. This, however, may have been the fruit of pragmatism rather than high moral principles. If an atom bomb had been within their scientific and technological grasp, it is by no means certain that they would have stood out against developing it.

No such issues arose on the Allied side, at least until Germany was defeated. The cussedness of human nature caused its problems, but there was a universal determination to win the war. If this called for an atom bomb, then an atom bomb there must be.

6 Resurrection of the British Project

When Halban and Kowarski arrived in England after the fall of France in June 1940, they were eagerly interviewed by British scientists and asked to write a full report on French fission research.

The previous three months had been busy ones for the British project. It had been almost moribund, but had now risen like a phoenix to vigorous life. The reason for this was a cogent three-page memorandum by Frisch and Rudolf Peierls, another German Jewish refugee with whom Frisch was then living, in which they argued the case for a bomb based on 'the use of nearly pure ^{235}U'. This was submitted in March 1940. Among the salient points were:

- Five kilograms of ^{235}U might suffice for a bomb, which might release as much energy as several thousand tons of dynamite.
- The uranium isotopes might be separated on a large scale by gaseous thermal diffusion using uranium hexafluoride. A hundred thousand separation units might be needed.
- The radioactivity produced in the explosion would constitute a further danger to life.

The memorandum generated a powerful momentum that was later to be transmitted across the Atlantic. Without it, the American bombs might well not have been ready in time to deliver the *coup de grâce* to Japan that was to end the war. Yet it is strange to reflect that little in the document would have struck Heisenberg and his colleagues as novel. They too were thinking of a ^{235}U bomb depending on fast neutrons, and Harteck had already been working for some months on the separation of the uranium isotopes by the very process now advocated by Frisch and Peierls. What was different was the tone of the approach. Whereas Heisenberg had covered the science comprehensively in his report to the German War Office, Frisch and Peierls' memorandum was directed in a businesslike way at a definite target.

Frisch and Peierls' ideas meant a complete change of direction for the British project. The discouraging attempts at reactor building could be laid aside, and effort concentrated on uranium isotope separation and bomb design. No longer need such a course be inhibited by fears that an impossibly large amount of ^{235}U would be required, or that the separation of the isotopes would be quite impracticable.

In April 1940 there was a further stimulus in the form of a visit by Allier with news not only of French progress but also of the ominous German interest in heavy water.

Soon there was a thriving programme under a strong committee chaired by Thomson and including Chadwick, Cockcroft, and other eminent scientists. It bore the curious name of the MAUD Committee. This arose because Bohr had cabled Frisch when Denmark was overrun, concluding with the words, 'Tell Cockcroft and Maud Ray Kent'. Not knowing that Maud Ray had been the Bohr children's governess and lived in Kent, Frisch thought the message was an anagram of 'radyum taken', meaning that the Nazis had confiscated the Copenhagen stock of radium. The name came to his mind when a non-informative title was needed shortly afterwards for the committee. Subsequent ingenuity produced the mythical interpretation *M*ilitary *A*pplications of *U*ranium *D*etonation.

The MAUD Committee held its first meeting on 10 April 1940 and soon presided over a well-balanced theoretical and experimental programme. There were four teams centred on the universities of Birmingham, Cambridge, Liverpool, and Oxford. Frisch joined Chadwick in Liverpool to measure basic nuclear properties, using the University's cyclotron, and further nuclear physics studies were made in Cambridge. Peierls in Birmingham continued to work on the problems of a ^{235}U bomb. Another refugee, Francis Simon in Oxford, developed the ideas for a uranium isotope separation plant. William Haworth, a Birmingham chemist, investigated methods of preparing uranium metal and uranium hexafluoride. Imperial Chemical Industries (ICI) and Metropolitan-Vickers also came into the picture.

In this setting, Halban and Kowarski's arrival posed a certain problem. Naturally they wanted to continue the work they had had to abandon in France, making use of their precious heavy water, but this appeared irrelevant to the revitalized British project, and there was some discussion as to whether it could be supported in wartime. Nevertheless they were eventually installed in the Cavendish Laboratory in Cambridge.

Here they had a large aluminium sphere built, which they filled with a uranium oxide/heavy water slurry. To keep the slurry well mixed, the sphere could be rotated. As in their previous work they inserted a neutron source into the centre of the sphere and made neutron intensity measurements at various points. On 16 December 1940 they claimed a k-value of 1.06. If correct, this value implied a potentially self-sustaining chain reaction, making Halban and Kowarski the first in the world to demonstrate such a possibility. Many felt, however, that the proof was inconclusive; there was too large a range of uncertainty in the measurements.

The logical way to settle the point was to make a bigger sphere with a greater quantity of uranium and heavy water, but with no hope of more heavy water the experimenters were stuck. They wanted to switch to graphite as moderator, but the British Government was unwilling to divert resources to manufacturing tons of ultra-pure graphite, and in any case Fermi was now active again in this field in the US and the work could be left to him.

Halban did not give up. He still, as a colleague said, 'had a single goal: to direct the team which would be the first to create a divergent chain reaction'. After lengthy negotiations it was agreed that his group at the Cavendish should move to Montreal in Canada to continue their efforts, but this did not take place until early 1943, and before they were ready to make a start Fermi's team had won the race. There was still a hope of making the first heavy water reactor, but even this was dashed when the US clamped down on co-operation. Despite this disappointment, the Montreal laboratory did eventually come to play an important part in the development of post-war nuclear projects in Canada, the UK, and France.

Although Halban and Kowarski's work had at first been regarded by the MAUD Committee as of doubtful relevance, a discovery reported in the US at the very time of their arrival in England was soon to put a new complexion on it. The discovery in question was made by Edwin M. McMillan and Philip H. Abelson, working at the University of California in Berkeley, across the bay from San Francisco.

In June 1940 they published a letter in the *Physical Review* announcing the discovery of neptunium, a new chemical element, among the products of bombarding uranium with neutrons. Nearly all the other species produced in this way had turned out to be fission products, though Fermi and others had mistaken them for transuranium elements, beyond uranium, which up till then had terminated the series of known elements. Now for the first time the existence of a genuine transuranium element had been established. About a dozen more such elements were to be discovered in the years to come, many of them again at Berkeley.

From the nuclear properties of the neptunium isotope they had identified, denoted by the symbol ^{239}Np, McMillan and Abelson knew that it must decay to the next transuranium element, which we now call plutonium, but this they could not detect. In their letter they suggested that the reason might be that the plutonium isotope concerned (^{239}Pu) is very long-lived; if so it would be much less radioactive than many of the other species present, making it much harder to detect. They were correct in their speculation; ^{239}Pu has a half-life as long as 24,390 years. (Half of it undergoes radioactive decay in this period.)

Any competent nuclear scientist who had read the 1939 paper on fission by Bohr and Wheeler could now see that ^{239}Pu was a potential bomb material. Chadwick arranged for an official protest to go to the US for publishing so suggestive a piece of information.

In Cambridge one of the MAUD team, Egon Bretscher, a Swiss, attempted to follow up McMillan and Abelson's discovery. He devised a chemical method for separating neptunium from irradiated uranium, with a view to allowing the neptunium to decay to plutonium. This illustrates a very important advantage of ^{239}Pu over ^{235}U: its isolation depends on separating two different chemical elements, not two different isotopes, and this is an everyday problem for the chemist. However the neutron sources available to Bretscher were too weak for useful results.

Another who read the McMillan and Abelson letter was Weizsäcker in Germany. Apparently he enjoyed shocking fellow-travellers on the Berlin underground in wartime by poring over English-language publications. Even before he saw the letter, he had speculated on the possibility of ^{239}Np being a nuclear explosive, and it had struck him too that an isotope separation might be avoided. Now he switched his ideas to ^{239}Pu. American information thus gave him a useful clue, but the Germans were never in a position to exploit it.

These were the first glimmerings of the plutonium alternative. Plutonium is now the explosive in all fission bombs, and its use was developed by the Americans for the Nagasaki bomb. To make it in quantity requires a nuclear reactor, such as Halban and Kowarski were trying to construct. In 1940, however, the existence of the element was unproved and its properties were of course unknown, so ^{235}U remained the first priority in British thinking.

Despite the excursions at Cambridge outside its original limits, the MAUD programme as a whole was compact and well-integrated. The objective was narrow and clear-cut: to determine whether an atom bomb could be made during the current war. The resources involved were small; only a handful of men were engaged at each university, and their expenditure was minute. Each group had its job to do, and understood how it fitted without overlapping into the total picture. In many cases the men already knew one another; they kept in touch and followed one another's progress, much as they were accustomed to do in peacetime.

A high proportion, including Frisch, Peierls, and Simon, were refugees, either aliens or naturalized British subjects. One reason for recruiting refugee scientists was simply that nearly all British-born scientists had by then been absorbed in other war work. It involved no security risk *vis-à-vis* Germany; on the contrary, the refugees constituted a powerful driving-force in their determination to forestall the Nazis. Thomson later wrote: 'It is noteworthy, and I hope will be noted by future Dictators, how dominating was the part played by physicists who had fled from Fascism and Nazism.'

The remarkable success of the MAUD Committee during its fifteen months' life, despite the blitz, the threat of invasion, and the other heavy commitments of its senior members, may be ascribed to the unity generated by the single-minded pursuit of its well-defined purpose. Everyone felt that the work must be done, and in a country fighting for survival against a tyranny, there was little hesitation on moral grounds.

The Committee finally produced two reports, which went to the Ministry of Aircraft Production on 29 July 1941:

- Use of uranium for a bomb.
- Use of uranium as a source of power.

The former was the more important. After an introductory sentence, it opened with the words:

> We should like to emphasize that we entered the project with more scepticism than belief though we felt it was a matter which had to be investigated. As we proceeded we became more and more convinced that release of atomic energy on a large scale is possible and that conditions can be chosen which would make it a very powerful weapon of war. We have now reached the

> conclusion that it will be possible to make an effective uranium bomb which, containing some 25 lb. of active material, would be equivalent as regards destructive effect to 1,800 tons of TNT and would also release large quantities of radioactive substances, which would make places near to where the bomb exploded dangerous to human life for a long period.

The report went on to say that material for three bombs a month might be prepared in a £5,000,000 plant (for uranium isotope separation) and that the first bomb might be ready at the end of 1943.

These quantitative statements were backed up by detailed information later in the report. There were sections dealing with the basic principle of the bomb, and the properties of ^{235}U that made it possible; the method of construction; the quantity of ^{235}U required; the damage the bomb would cause, both by explosion and by radioactive contamination; the method of separation of the uranium isotopes, and the plant required for the purpose; and the preparation of uranium hexafluoride for the isotope separation plant.

The minimum amount of ^{235}U for a self-propagating chain reaction was calculated to lie between about five and forty-three kilograms, depending on what was assumed about neutron production and losses. Perhaps twice as much would be needed for an efficient explosion, but the quantity might be greatly reduced by means of a tamper, such as a thick layer of steel, which would reflect some of the emerging neutrons back into the uranium.

Taking all this into consideration, the MAUD report adopted for discussion purposes a total mass of ten kilograms for a bomb. This would be divided into two separate masses of five kilograms each, too small to explode by themselves, but able to do so when brought together. The combined mass, it was believed, would explode spontaneously, because there are always stray neutrons around to trigger off the chain reaction. To avoid a mere fizzle, the two parts must be brought together very rapidly, and the report suggested firing them at one another from opposite ends of a gun barrel, giving a relative speed of 6,000 feet per second (about 4,000 mph) on impact. It was considered that the explosion of 2 per cent of the ^{235}U should be readily achieved by this 'gun method', causing the same damage as 1,800 tons of TNT (Fig. 10).

It is interesting to compare these early estimates with the Hiroshima bomb, which used about sixty kilograms of ^{235}U,

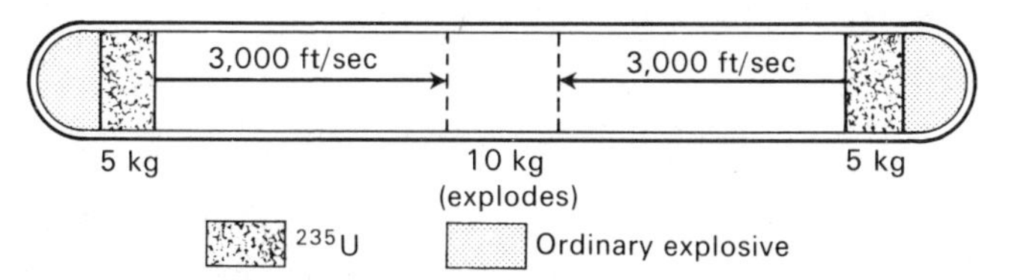

Fig. 10. The MAUD report concept of an atom bomb.

Ordinary explosives are used to drive the two 5 kg lumps of ^{235}U together very rapidly inside a gun barrel, which is closed at both ends. In the Hiroshima bomb only one of the ^{235}U lumps moved, and it was much smaller than the stationary lump; the two came together to form a spherical mass of ^{235}U.

detonated by the gun method, and achieved an efficiency of about one per cent.

Of some concern in the MAUD report was the fact that ^{235}U is itself a weak source of neutrons, which might trigger the explosion prematurely. The neutrons come from the occasional spontaneous fission of ^{235}U nuclei, in which the nucleus breaks up without any external stimulus such as the capture of a neutron. This phenomenon makes it all the more necessary to assemble the explosive mass very rapidly.

To separate the uranium isotopes it was proposed to use diffusion of uranium hexafluoride gas through 'gauzes of very fine mesh', otherwise referred to as 'membranes' or 'porous barriers'. This is a technique with a long history. Lord Rayleigh stated the basic principle in 1896, and the first ever isotope separation, of the neon isotopes, was carried out with its aid by Aston in 1913. Aston only obtained a slight separation, but later workers, notably Gustav Hertz in the 1920s, used it with considerable success.

The essential feature is simply that the lighter isotope can travel slightly more easily through the membrane than the heavier. As uranium hexafluoride traverses a membrane, therefore, the gas coming through is enriched to a small degree in the lighter ^{235}U isotope, while that remaining behind is depleted. The effect during a single passage is very small, so it is necessary to repeat the process many times with large numbers of membranes.

Frisch and Peierls selected this technique as a result of a study in which they designed imaginary large-scale plants utilizing different separation methods. They abandoned their original proposal of gaseous thermal diffusion for various reasons: it was slow, it consumed large amounts of power, it required temperatures that might decompose the uranium hexafluoride, as indeed the Germans discovered, and laboratory experience was discouraging. They also rejected centrifuging (which operates like a cream separator), considering it promising in principle, but requiring too much precision engineering.

Although the basic idea of membrane diffusion is simple, its elaboration to give a large-scale plant is complicated. This task was undertaken with great skill by Simon and his colleagues in Oxford. A short account of their work appears as an appendix to one of the MAUD reports.

The MAUD reports disappeared into the Government machine, while many of those who had created them wondered what was going on. In fact the reports were the subject of intense discussion. The prize was tremendous, but did the chances of success justify the necessary investment of national resources in wartime? So far as nuclear power was concerned, the answer was unequivocally negative. As regards the bomb, two points crystallized out: first, that the report appeared over-optimistic, and its ideas needed firming up, and secondly, that Canada or the United States might be a better location than Britain for a uranium isotope separation plant. On the second point a very important consideration was the vulnerability of the plant to air attack, since it would cover many acres, use large amounts of electric power, and have to operate over long periods without interruption.

The MAUD Committee were all academics, and for the next phase men from industry had to be brought in. A new body, again with a name that gave nothing away, the Directorate of Tube Alloys, was established. This had its home in the Department of Scientific and Industrial Research, and Wallace Akers was introduced from ICI to take charge. Other ICI men were also included. Some of those associated with the MAUD Committee (Chadwick, Simon, Halban, Peierls) were appointed to the Tube Alloys Technical Committee, but others were left in the dark for months. Oliphant was indignant about the reorganization: 'I can see no reason whatever why the people put in charge of this work should be commercial representatives completely ignorant of the essential nuclear physics upon which the whole thing is based.' There were to be similar scientific *cris de cœur* in the US when the inevitable happened, and the big military and industrial battalions took over. Akers fortunately had the right personality and abilities, and the Tube Alloys scientists were gradually won over.

7 Launching of the American Project

In mid-1941 the MAUD reports went to the US, and the focus of interest began to shift across the Atlantic, although work continued in Britain.

American concern had been reawakened in the spring of 1940 by snippets of news from Britain, France, and Germany. One result was that Briggs's Advisory Committee on Uranium gave cautious support to further work on the uranium/graphite system, and in May of that year Fermi and Szilard were able to report an encouragingly low value for the absorption of slow neutrons in graphite. This was the start of a programme aimed at building the world's first man-made nuclear reactor.

The pace of this work was dictated to quite a degree by the availability of materials, and their purity. Some impurities, such as boron, absorb slow neutrons so strongly that they can only be tolerated at a level of a few parts per million. To start with, neither the quantity nor the quality of the uranium and graphite was anything like sufficient for a self-sustaining chain reaction, so Fermi and his colleagues concentrated on small-scale experiments to collect basic data on their materials. Larger-scale work had to wait for over a year.

The discovery of neptunium at Berkeley prompted a second line of research, the investigation of plutonium. This was taken up by Glenn T. Seaborg in December 1940 with the aid of Berkeley's powerful cyclotron. Even with this instrument the quantities of plutonium he could produce were minute in the extreme, measured in micrograms (millionths of a gram) compared with the tons that exist today. To identify and investigate it, he and his team had to use very sensitive radiochemical techniques, many of which had been invented by Hevesy thirty years before in Rutherford's laboratory. The discovery of the new element was announced in January 1941, and the following May came the vital information that like uranium it can readily be induced to undergo fission.

A further stimulant to American thinking was the separation of minute amounts of the uranium isotopes in a mass spectrometer, and the demonstration that the one responsible for the fission observed with slow neutrons was indeed ^{235}U, as Bohr had surmised. This result was due to John R. Dunning, a colleague of Fermi's at Columbia University, and it aroused much interest at an American Physical Society meeting in April 1940. It led to efforts, mainly at universities, to separate the uranium isotopes on a larger scale.

Such work was carried out independently of that in the UK, Dunning himself in co-operation with Urey covering some of the same ground as the MAUD Committee. They followed the British in taking up the method of gaseous diffusion through a membrane, as well as in rejecting gaseous thermal diffusion.

The Americans, however, put much greater emphasis on the centrifuge method at this stage. Jesse W. Beams at the University of Virginia had used it successfully for chlorine, and he now sought to extend its application to uranium. There was also a novel method due to Abelson, one of the discoverers of neptunium, who had the idea of carrying out thermal diffusion with liquid instead of gaseous uranium hexafluoride; this greatly improves the efficiency, though by no means up to that of membrane diffusion. Towards the end of 1941, yet another novel method, electromagnetic isotope separation, was introduced, as described below.

Fermi's, Seaborg's, Dunning's, and Beams's work, though not Abelson's, all came within the purview of the Briggs committee. There was also another item on Briggs's agenda: production of heavy water as an alternative moderator, in case Fermi's graphite proved a failure. This was inspired by news of Halban and Kowarski's experiments in the UK on a uranium/heavy water assembly, and it was dear to the heart of Urey, the discoverer of heavy water.

So far the American nuclear programme had been carried on very much in the spirit of scientific curiosity and individual initiative characteristic of university research, and it was co-ordinated as much by the grapevine as formally by the Briggs committee. Up to a point, this casual approach was successful. The lines of attack just mentioned proved to be the very ingredients needed for developing large-scale production of the two principal nuclear explosives, ^{235}U and ^{239}Pu. Uranium isotope separation yielded the former, while the latter was to be produced in reactors such as Fermi was trying to build, and recovered from them by methods based on Seaborg's chemical studies.

There was, nevertheless, a growing feeling that more ought to be happening. Nuclear physicists wondered why they were not being enlisted in official programmes. Those actually involved chafed at the tardy provision of funds.

Among those who were dissatisfied was Vannevar Bush, President of a private research organization, the Carnegie Institute. He was an inventive electrical engineer, unashamedly proud of his country and its moral heritage, threatened now by Nazism and Fascism. Although the US was still at peace, he began to devote himself to mobilizing American science for war, and persuaded President Roosevelt to set up a National Defense Research Council with himself as Chairman. Closely associated with him was James B. Conant, a chemist who had become President of Harvard.

In April 1941 Bush asked the National Academy of Sciences to review the whole nuclear programme. Chairing the reviewing committee was Arthur H. Compton, who had been awarded the Nobel prize for his discoveries in nuclear physics. He needed little prompting, being already worried that progress seemed so slow.

Another who sought to ginger things up was Lawrence, the inventor of the cyclotron. It was in his laboratory at Berkeley and with the aid of one of his instruments that neptunium and plutonium had recently been discovered, and of course this interested him greatly. Conscious of the darkening war situation and of the restiveness among nuclear physicists, he began to press for official action, and he also toyed with the idea of converting his smaller, partly redundant cyclotron into a kind of super version of the mass spectrograph, the instrument invented by Aston for very small-scale isotope separation (Chapter 1).

Bush made another move in June when he obtained Presidential sanction to set up an office with full-time staff to co-ordinate military research, with himself as Director. Briggs's committee was brought under the new Office of Scientific Research and Development, reporting to Conant and cryptically renamed 'S-1'.

In the midst of these struggles to get things going, the draft MAUD reports fell into Bush's and Conant's laps in July 1941. The timing could not have been better if it had been engineered. Here was the note of conviction, the sense of solid ground, that the Americans needed. The official history of the US Atomic Energy Commission describes July 1941 as 'the turning point in the American atomic energy effort' and states that 'news from Britain', meaning especially the MAUD reports, was the most important factor 'in effecting a new approach'. Another official American account speaks of exchanges with the British bringing 'a sense of urgency'.

The natural outcome might have been a joint project, with the Americans and the British participating as equals, but there were hesitations on the British side. America was not yet a belligerent, so what about secrecy? Might Britain lose control of a decisive weapon? What about the post-war situation? So the opportunity was lost, with traumatic consequences for the British later on. Nevertheless there was a honeymoon period with many contacts between the two countries and full sharing of information.

One of these contacts was between Lawrence and his old friend Oliphant, who paid him a visit during the summer of 1941 and gave him an account of the MAUD work. Lawrence was much impressed and rang up Arthur Compton, who then invited him to his home in Chicago to meet Conant to discuss the prospects for atom bombs. Conant played hard-to-convince and then suddenly challenged him, 'Does this seem so vital to you that you are prepared to devote to it the next years of your life?' Lawrence was momentarily taken aback; there were many things he wanted to do, which he would have to put aside. Nevertheless, he replied, 'If you tell me this is my job, I'll do it.'

The immediate practical step was to modify his small cyclotron, and in November he got together some of his best men at Berkeley to make the necessary changes. Only a natural optimist with a flair for getting scientific machines to work would have tackled the job so quickly and boldly, considering that a new and untried principle was involved. Previously everyone had thought that if you attempted to put large quantities of material through a mass spectrograph, in order to obtain sizeable amounts of separated isotopes, you would be bound to fail. This was because the atoms of uranium or whatever passing through the instrument all carry the same electric charge, and therefore repel one another. The more material you try to pass, the closer together are the atoms, and the more important do the repulsions become. The result is that instead of clean, sharp beams of atoms of each isotope, you get fuzzy beams, which eventually overlap so much that there is no separation of the isotopes. Lawrence's hunch was that this unfortunate effect might be neutralized by feeding in particles of the opposite charge, but it was by no means self-evident that this expedient would succeed.

Lawrence had the device working by 2 December 1941 and found that his hunch was right. By the following February he was turning out small specimens of the uranium isotopes for nuclear physical measurements. The new instrument was the forerunner of the 'calutrons' used to make material for the Hiroshima bomb, and of the electromagnetic isotope separators in use today.

Electromagnetic separation is in some ways the odd one out among the four uranium isotope separation methods that were by this time under investigation in the US. The others all had perforce to use uranium in the form of the corrosive hexafluoride, and the difficulties of working with this nasty material resulted in slow progress. Furthermore the electromagnetic method was the only one that could give a more or less complete separation in a single stage of operation. In the other three, each stage achieved only a slight separation which had then to be multiplied up through very large numbers of stages, as Simon had discussed in one of the MAUD reports. This meant that the plant must consist of tens of thousands of similar units. If the units had been simple, straightforward ones, there would have been no particular difficulty, but they were novel, often required accurate machining (especially the centrifuges), and had to stand up to uranium hexafluoride. Unfortunately the electromagnetic method also called for numerous complex units, though for a different reason. Each unit was able to handle very much more material than a mass spectrograph, but the quantity was still small compared with what was needed for a bomb, and there was nothing for it but to build a lot of units.

Lawrence's first success with the electromagnetic method came just before the disastrous Japanese attack on Pearl Harbour on 7 December 1941, which brought the US into the war and gave the making of atom bombs a new urgency. Shortly afterwards Bush appointed the three Nobel laureates, Compton, Lawrence, and Urey, as programme chiefs. For Lawrence and Urey this served mainly to establish their status in the tasks they were already undertaking, but for Compton it meant a major new responsibility: plutonium, an element nobody had yet seen, but which would be needed by the kilogram.

By this time Fermi's reactor work was recognized as an early phase of a plutonium bomb project, since it was only in nuclear reactors that plutonium might be made in bulk. Compton had to cover this as well as all the later phases of the project, right up to making the bomb itself. Throughout 1942 he was in charge of the entire plutonium programme.

He described the task as 'a heroic act of faith'. Faith was something he understood from his deeply Christian family background, and that may also have been the source of his readiness to go all out for an important goal.

The relevant effort was spread right across the US, and the need for secrecy made communications difficult, so Compton decided to centralize in Chicago, in a so-called Metallurgical Laboratory. This was a deliberately vague name, abbreviated colloquially to Met. Lab.

Right at the start, in January 1942, he announced a time-schedule:

- By July 1942, to determine whether a chain reaction is possible. (July 1942)
- By January 1943, to achieve the first chain reaction. (December 1942)
- By January 1944, to extract the first plutonium from uranium. (December 1943)
- By January 1945, to have a bomb. (July 1945)

The dates in brackets are those actually achieved; they are a remarkable witness to American drive and determination. The first two items depended on the work under Fermi, and the third on that under Seaborg, while work on the fourth item had still to be activated when Compton took over.

Seaborg's work assumed Fermi's success. The main task was to devise a method of recovering the tiny proportion of plutonium that would, it was believed, be produced in the uranium in Fermi's hoped-for reactor. This necessitated fairly extensive studies of plutonium chemistry, which proved to be very interesting – quite different from what had been expected, and quite complicated.

By early 1944 there would be grams of plutonium to work with, prepared in reactors, but in 1942 the cyclotrons could only yield micrograms. This meant that most of the time it was mixed with far larger quantities of other substances, and its presence could only be detected by its radioactivity. Two of the Met. Lab. chemists were able to make a single microgram of pure material in August 1942, like a speck of dust just visible to the naked eye on the wall of a small tube, but this was a considerable *tour de force*; generally the plutonium was invisible. Yet with these minute traces a separation process had to be developed that could be used in a chemical plant handling tons of uranium.

Another facet of the work was the investigation of the products of the fission process. It was the identification of certain of these species that had led to the discovery of fission in the first place, and increasing numbers of them were being discovered – sixty-four by

May 1942. Plutonium would have to be separated from these as well as from uranium; they were radioactive and if left with the plutonium would make it much more difficult to handle.

The specialized techniques required for all this research were unfamiliar to most chemists at that time, and many of Seaborg's team had to learn while doing the job. In the summer of 1942 it was therefore a useful bonus to have the young French radiochemist, Goldschmidt, at the Met. Lab. He had worked in the field since 1933, first with Marie Curie and later with Joliot. Halban, trying to build up the Montreal laboratory, arranged for him to spend a week in Chicago to learn about Seaborg's work before coming to Montreal, and it proved so mutually beneficial that he stayed for three months.

As the various projects under Compton, Lawrence, and Urey progressed in the early months of 1942, there was a growing sense that fateful decisions must soon be made. The work would have to move from the laboratory to the factory if nuclear explosives were to be prepared in bomb quantities. Expenditure, which had been running at hundreds of thousands of dollars, might jump to as many hundreds of millions of dollars, yet success was far from assured. Did the war situation justify such a large diversion of national resources?

The day of reckoning was 23 May 1942, when Conant called together the S-1 leaders to his office in Washington.

They had before them two possible explosives: ^{235}U and ^{239}Pu. They had three uranium isotope separation methods that might produce ^{235}U: gaseous diffusion, centrifuging, and electromagnetic separation, though they seem to have ignored or overlooked Abelson's liquid thermal diffusion. They had two possible types of nuclear reactor that might manufacture ^{239}Pu: uranium/graphite and uranium/heavy water. Five routes in all to a bomb.

The decisive argument that day was that the Germans probably had a two-year lead, and with five possible routes might already be well ahead with one of them. If they got an atom bomb first, even the mighty United States might be defeated.

With this reading of the situation the S-1 Committee wanted not necessarily the best, but the fastest route. There was, however, so much uncertainty about all their candidates that they were unable to choose, and they made the 'Napoleonic decision' (Conant's phrase) to recommend development of all five. This was accepted by Bush and by the Ministry of War, and the die was cast.

The decision was momentous, fantastic. It meant a commitment

Table 5. Routes to nuclear explosives considered in the US in 1942. All except route 4 were considered by the S-1 Committee in May 1942, but routes 3 and 6 had been abandoned by the end of 1942. Route 4 was developed by the US Navy and ultimately used by the Manhattan Project.

Explosive	**Route**
^{235}U	1. Gaseous diffusion.
	2. Electromagnetic separation.
	3. Centrifugation.
	4. Liquid thermal diffusion.
^{239}Pu	5. Uranium/graphite reactor.
	6. Uranium/heavy water reactor.

to a diffusion plant before the essential diffusion membranes or barriers were available; to a centrifuge plant when even laboratory success was minimal; to an electromagnetic plant whose principle had only been tested on a microgram scale; and to nuclear reactors whose very possibility had yet to be demonstrated. In peacetime it would have been thought foolhardy, to say the least, to build on such insecure foundations.

The five routes were effectively reduced to three before the end of 1942. The centrifuge project was dropped for the same reason as had been given in the MAUD reports: too much precision engineering. The high-speed centrifuges must be very accurately balanced, because the least wobble can develop into a ruinous instability. Beams had been able to construct a small laboratory instrument and to pass some uranium hexafluoride through it, obtaining a very slight enrichment of the ^{235}U by early 1942. Even this minor success was less than had been predicted theoretically, while the scaling-up difficulties appeared tremendous; tens of thousands of large units would have to be built to very fine tolerances, and would have to operate with scarcely a breakdown.

The other route that was dropped was the one based on the uranium/heavy water reactor. This was partly because Fermi's uranium/graphite work was progressing so well, and partly because making enough heavy water for plutonium production reactors would take a considerable time. Action was nevertheless taken on heavy water supplies by way of insurance.

For the production phase of the nuclear programme, as Bush had foreseen well in advance, skills and experience quite different from those of the research teams would be required and he had the Army in mind. Now he and Conant turned to the US Army Corps of

Engineers. There was a gigantic construction task ahead, greater than anyone yet fully appreciated, for which men who had had to build large training camps and air bases were well suited.

It would also be necessary to bring in industrial firms as contractors to the Army, and this caused a violent reaction among the scientists at the Met. Lab. There was a show-down at a meeting there in June 1942, when Compton faced, in his own words, 'a near rebellion'. He opened the meeting by reading the story of Gideon from the Old Testament, implying that he would prefer a small fully-committed team to a large lukewarm one.

There is disagreement over exactly what was at issue.* One of the American scientists present says that their objection was to the proposed choice of Stone and Webster, the Army's regular contractors, for the plutonium production plants; the scientists thought they were the wrong sort of firm. On the other hand, the refugees from Europe, and Wigner in particular, were suspicious of *all* big firms, including perhaps most of all the firm eventually chosen, the chemical giant, E. I. du Pont de Nemours. Compton says: 'It was very difficult at first, however, for Wigner to believe that anything good could come from co-operation with a great industrial organization such as du Pont.' Such companies, he had been taught in Europe, were the tyrants of the American democracy.' The introduction of du Pont in the autumn of 1942 was to rankle in Chicago for some time.

The Army selected Colonel Leslie R. Groves to take charge of the Manhattan Project, as the whole organization was now named, and promoted him to Brigadier General. Groves had been responsible for building the Pentagon in Washington, and that building is a monument to his ability to get things done through a large and complex organization. He had a colossal capacity for work, drove himself and others hard, and had no patience with inefficiency. In running the Manhattan Project he was to be here, there, and everywhere, wherever there was a bottleneck or a key decision to be made, yet he never lost sight of the main objectives. It is arguable that but for Groves the atom bombs might not have been ready before the end of the war.

To him making the bombs was a job to be done, a job on which the outcome of the war might depend. The bombs themselves were weapons, and the Army were the weapons experts. Secrecy was

*Conflicting accounts are given by Compton in *Atomic Quest* and Libby in *The Uranium People* (see Further Reading); the latter places the event in the autumn of 1942.

essential, and must be rigorous. His approach was simple, direct, unsubtle, paying little attention to wider issues.

He came to see Bush on 17 September 1942. Bush had not been consulted or even informed of Groves's appointment, and his initial reaction was cool. Groves himself said, '[Bush] felt I was too aggressive and might have difficulty with the scientific people'. Before long, however, the two became firm friends. By and large the scientists accepted the Army and the Army's security restrictions as inevitable, though they grumbled and sometimes found Groves little short of intolerable.

For the British the advent of the US Army was disastrous, because it led to an almost total embargo on the flow of American information. This was a bitter pill, considering the value of the MAUD work to the Americans. The embargo was not lifted until nearly a year later, when Churchill and Roosevelt negotiated the so-called Quebec Agreement. Even then the British were excluded from several key areas, including much of the Met. Lab. work.

The embargo was particularly damaging to the Montreal laboratory, which was planned on the assumption of American co-operation, and was ready to start operating early in 1943 just after this had been brusquely cut off. Halban had had a uranium/heavy water project in mind as a centre-piece for the laboratory, but his ambitions were frustrated by diversion of most of the available heavy water to Chicago. A little alleviation of the situation was achieved by private visits to Chicago in February by Goldschmidt and another Frenchman, Pierre Auger. They still had their Met. Lab. badges, so they simply walked in and were welcomed by their former colleagues, some of whom were themselves smarting under the new management of the Manhattan Project. The Frenchmen returned to Montreal with valuable booty, including small specimens of plutonium and fission products that Goldschmidt had worked on the previous summer.

Under Groves meanwhile the scale and character of the American programme were being rapidly transformed. People were employed by the tens of thousands on crash programmes to produce ^{235}U and ^{239}Pu. To a large extent the story becomes one of huge projects carried out by industrial contractors, who often tended to overshadow the scientists. Yet the scientists still had many vital contributions to make, and it is these that will be emphasized in the following chapters.

8 Separation of the Uranium Isotopes

One of General Groves's first moves was to acquire a large tract of land in Tennessee which had been chosen as a production site. There a gaseous diffusion plant (K-25) and an electromagnetic separation plant (Y-12), as well as a large power station, were built by industrial firms under Army contracts, together with a whole new township and research laboratories, which were later to become the world-famous Oak Ridge National Laboratory (ORNL). The plants were seven miles apart in separate valleys so that an accident to one would not affect the other.

The usual practice with a new process is to test it out first on a reduced scale in a pilot plant, and only then to proceed to the full-scale plant. Unforeseen snags are brought to light and can be avoided in the final design. The plants Groves had to build were so novel that in normal times this procedure would have been regarded as doubly necessary. But then the supposed race with the Nazis might have been lost.

The decision was taken, first for the gaseous diffusion plant, and then for the electromagnetic plant, to skip the pilot plant. Groves said that he was prepared to cut corners in this way partly on account of his great confidence in Lawrence's ability and drive. Nevertheless it was a recipe for trouble, and trouble developed in plenty.

The electromagnetic plant, Y-12, was the first to go ahead. Construction started in February 1943 and the first unit was completed by August. The research for it was carried out in Lawrence's laboratory in Berkeley, and the scientists there were very closely integrated with the firms involved. Fifty of them were actually transferred to Tennessee Eastman, who had to operate the plant.

The Y-12 plant had to take natural uranium, containing only 0.71 per cent of the desired ^{235}U isotope, and obtain from it almost pure ^{235}U at something over 90 per cent, which was what the bomb designers wanted. Lawrence originally hoped to achieve this in a single stage, but to Groves this was over-optimistic, and Y-12 was planned as a two-stage system. The first stage, alpha, would give an initial enrichment up to about 15 per cent, and the second stage, beta, would complete the process. The roughly twentyfold increase in concentration in the alpha units meant that about 95 per cent of the unwanted ^{235}U isotope had been removed, and only about 5 per cent remained (Fig. 11), so the beta units would have far less uranium to handle than the alpha units, and would be much smaller. Eventually there were nine alpha units ('racetracks' as they were called from their general appearance), whose product could be fed to eight beta units.

The essential process took place in devices called 'calutrons', in which uranium atoms traversed a high vacuum under the influence of electric and magnetic fields which sorted out the ^{235}U and ^{238}U atoms. The separated material was then collected in, for instance, special cavities, from which it was removed from time to time.

Some idea of the size and complexity of the plant can be gained from the fact that it required 22,000 operators. Each alpha racetrack comprised no fewer than 96 tanks containing the calutrons, and was 122 feet long, 77 feet wide, and 15 feet high (Plate 13). Never before had anyone created such vast, highly evacuated spaces, while the enormous electromagnets threatened to swallow up nearly 100,000 tons of copper, to the detriment of other wartime projects. To overcome the latter difficulty, silver was used as a substitute; 86,000 tons were borrowed from the US Treasury, and returned with only minor losses after the war.

In one sense the process was very inefficient: only about 10 per cent of the uranium atoms injected into the calutrons reached the collectors. The rest was spread all over the apparatus in the tanks,

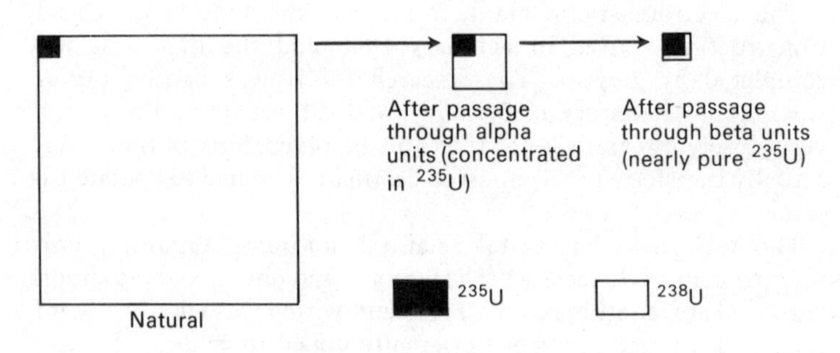

Fig. 11. Composition of uranium in electromagnetic separation.

and to retrieve it involved a troublesome job of dismantling and 'pickling' the pieces of equipment in acid.

This required a large chemical plant. There were other chemical operations, too, to prepare the special form of uranium needed as a feed to the calutrons (uranium tetrachloride), and to purify and process the separated isotopes from the collectors. At first the Berkeley physicists paid only scant attention to the chemistry, but the decision to adopt a two-stage enrichment scheme for Y-12 brought its importance and difficulties home to them. This was because the uranium scattered over the equipment in the beta units would be the valuable, partly enriched uranium prepared with so much trouble in the alpha units. It was vital to recover it and to put it through the beta units again.

After inspecting Y-12 in May 1943, Lawrence told his Berkeley colleagues: 'When you see the magnitude of that operation, it sobers you up and makes you realise that whether we want to or no, that we've got to make things go . . . It's going to be an awful job to get those racetracks into operation on schedule. We must do it.'

It certainly did prove an 'awful job', requiring all of Lawrence's enthusiasm to keep things moving and to sustain morale during the dark periods of disappointment.

When the first alpha unit was started up at the end of 1943, it began to run erratically after a few hours and then shut down altogether. There were shorts in the magnets. This was the first in a long series of equipment failures. It was particularly frustrating when some minor problem – on one occasion, a mouse – involved opening up a tank, doing the work required, and then tediously pumping down for many hours. Nevertheless the second alpha unit produced material for experimental work early in 1944, and gradually the 'bugs' were eliminated.

The scientists were in constant demand to diagnose and cure the problems; there was no question of their being slowly pushed aside by their industrial counterparts, as happened in some other areas. Moreover after the 1943 Quebec Agreement they were reinforced by some thirty-five scientists from the UK, who were made very welcome and were completely integrated into the project, moving freely between Berkeley and Oak Ridge. Oliphant, indeed, took charge when Lawrence was away from Berkeley.

Meanwhile the gaseous diffusion programme was struggling along. Dunning at Columbia University had already linked up with the M. W. Kellogg Company, with a view to development of the large-scale equipment that would be needed for a plant, and at the end of 1942 this firm was asked to take over design of the plant itself. A new subsidiary, the Kellex Company, under Percival C. Keith, was formed for the purpose. An operator was also required for the eventual plant, and the giant Union Carbide and Chemicals Corporation was brought in. The plant was to be a giant too; the building that housed it was the largest in the world.

It was one thing, however, to make these arrangements, and another for the firms concerned to get down to business while the design of the basic components remained so vague. Much of the effort in 1943 therefore went into a variety of test units, some quite large. Construction work for the plant, K-25, did not start until the middle of 1943, well behind Y-12.

The most obstinate problem was the production of the diffusion membranes or barriers, on which the whole process depends. The barrier must be porous, like an earthenware pot, and the thousands of millions of pores must be microscopic yet reasonably uniform in size; a large pore would in fact be a leak, allowing the uranium hexafluoride molecules through without differentiating between the isotopes. Yet the barrier must be robust, able to withstand the pressure required to force the gas through, and the pores must not become clogged as a result of corrosion. The scientists tested hundreds of materials, many specially fabricated, in attempts to meet the exceedingly tough combination of requirements. Millions of square feet were needed, yet for a long time the making of even a few square inches was chancy and unreliable. By the latter half of 1942 nickel had been identified as a likely barrier material, able to withstand corrosion by uranium hexafluoride, but how could it be fabricated into the required porous form?

Urey's laboratory at Columbia University, named the SAM (Substitute Alloy Materials) Laboratory from May 1943, wrestled with the problems. In contrast to the Berkeley scientists working on Y-12, Urey and his team tended to become backroom boys, isolated from the nitty-gritty problems of the firms who had to build and operate the production plants. Partly this was because there was a sharper dividing line between research and industrial action in this field, but partly too it was a matter of personalities. Lawrence at Y-12 exuded hope; Urey often became discouraged. Lawrence did whatever was needed to get the process to work; Urey focused on the science. Moreover, Urey ultimately fell out with Keith, the Kellex man, who should have been his ally.

By the autumn of 1943 no definite solution to the barrier problem had yet been obtained. At this point the Quebec

Agreement enabled Groves to call in the British, who had put much effort into their own gaseous diffusion project, though no industrial plant was yet in prospect in Britain. Peierls was attached as a consultant to Kellex; a British mission visited America in the winter of 1943/4; and several British scientists and engineers spent some months with the American project in 1944. Co-operation was, however, much less successful than in the case of the electromagnetic project. The British team had their own strongly-held views on what a gaseous diffusion plant ought to be like, but the Americans were already committed to their own design to the tune of tens of millions of dollars. The British approached K-25 with a certain academic detachment, whereas the Americans knew they urgently had to make it work, come what may. Nevertheless the Kellex engineers found value in going over all aspects of the plant with a new and knowledgeable group.

On the crucial question of the barrier, the British had no magic formula to offer. The Americans had two front runners, the 'Norris–Adler barrier', a kind of nickel mesh screen with vast numbers of minute holes, which had been the subject of eighteen months' work under Urey, and the 'Johnson barrier' developed relatively recently under Keith at Kellex, and made from nickel powder, sintered together to form a porous compact. A production plant for the Norris–Adler barrier was already under construction, but the membranes were still too fragile, difficult to weld, and variable in performance. The Johnson barrier was successful when made by hand, but no method of large-scale production had been devised.

Keith felt that it was flogging a dead horse to press on with the Norris–Adler barrier, which he disparagingly called a 'lace curtain'. Urey thought it madness to switch to a new and comparatively untried concept; if the Norris–Adler barrier could not be made to work, K-25 should be abandoned altogether. Urey became tense and disheartened, and all but resigned, compelling Groves to shift the main responsibility on to other shoulders.

Groves kept work going for as long as possible on both types of barrier – which incidentally satisfied neither protagonist – but by early 1944 a decision had to be taken. At a crucial meeting of all the parties, including the British, the Kellex engineers proposed ripping out all the machinery installed to manufacture the Norris–Adler barrier, and bringing in thousands of operatives to make the Johnson barriers by a simple manual process. The British were sceptical, and one of them said that 'something in the nature of a miracle' would be needed, but Groves accepted the Kellex plan, and the bold decision was justified by results. It could perhaps be regarded as a consolation prize for the British that a factory in Wales supplied much of the very high quality nickel powder that was needed.

Development of the barriers had absorbed the efforts of literally hundreds of scientists and production engineers. Until the problem was solved, the diffusion plant, K-25, was held up. Barrier tubes were finally forthcoming on a significant scale towards the end of 1944, and thereafter K-25 units came into action as fast as barriers were delivered. It was possible to feed the first few units with uranium hexafluoride on 20 January 1945.

Next to the barriers, pumps were the hardest components to design. They must circulate the corrosive uranium hexafluoride reliably at high speed over long periods, and they must be completely leak-tight, to keep out atmospheric moisture, yet there was no lubricant that could withstand the hexafluoride and be used to make a seal. A novel and satisfactory design was at last forthcoming in the spring of 1943.

When both K-25 and Y-12 fell behind schedule, the idea arose of linking them in tandem. They had been intended as alternatives, each capable of carrying out the entire enrichment process; if one failed, the other might succeed. The new proposal was to use the gaseous diffusion plant, K-25, for an initial, limited enrichment of the uranium, and to feed the resultant product to the electromagnetic plant, Y-12, which would complete the enrichment. Each plant would then be employed to the best advantage, K-25 for the major quantities of material which had to be handled at the outset, and Y-12 for the smaller quantities of partially enriched material. Moreover the final stages of K-25 could now be cancelled, so that less plant would have to be built.

The K-25/Y-12 combination was the white hope of 1943, but by early 1944 continued poor progress led to further discouragement. Then, however, yet another door opened. Abelson had been pursuing his work on liquid thermal diffusion, almost forgotten by the Manhattan Project. The method involved forcing liquid uranium hexafluoride into a vertical annular pipe, heated on the outside by steam and cooled on the inside by water. Just as in gaseous thermal diffusion, the lighter isotope then concentrated near the hot wall, where it tended to rise up the tube, and the heavier near the cold wall, where it tended to sink down the tube. The US Navy, being interested in nuclear power for submarines, supported Abelson's work, and in 1941 he moved from the National Bureau of Standards in Washington to the Naval Research Laboratory in Anacostia, partly for the sake of a good supply of steam for heating his tubes. By January 1944 the construction of a hundred-column plant had been started in the Philadelphia Navy Yard, and Abelson hoped to begin production of enriched material on a small but significant scale by July.

There was very little contact with the Manhattan Project, the Navy not being a party to its secrets. News of Abelson's work nevertheless reached Groves by a roundabout route in April 1944. His team made a quick study of the process, and on 27 June a contract was signed with the H. K. Ferguson Company for a liquid thermal diffusion plant at Oak Ridge, to be built in ninety days. The firm decided that in such a short time all they could do was to put up twenty-one exact copies of the plant in the Navy Yard. This became S-50. Enough of the plant was operating by October to make token deliveries of slightly enriched uranium to Y-12, more or less within the ninety-day deadline. The whole plant was on stream by the following March.

The S-50 plant only raised the proportion of ^{235}U from the 0.71 per cent of natural uranium to 0.86 per cent, which might seem a trivial increase. It does nevertheless imply that when the S-50 product replaces natural uranium as a feed to Y-12 or K-25, the amount of ^{235}U processed in the second plant rises by roughly 21 per cent; it is equivalent to a 21 per cent increase in capacity. This was significant when the project was counting the kilograms towards the first ^{235}U bomb.

So Groves now had three strings to his bow, all in varying degrees incomplete and unreliable, but all steadily improving. The problem in 1945 was how to combine them to the best advantage.

At first, the S-50 product was fed direct to Y-12. Then on 12 March, when K-25 had demonstrated that it could produce material of 1.1 per cent enrichment, it was interpolated in the chain, which thus ran from S-50 to K-25 to the alpha racetracks of Y-12 and then the beta racetracks. Another landmark was reached on 10 June, when K-25 was able to produce 7 per cent material; this was of a sufficient degree of enrichment for feeding direct to the beta units of Y-12, bypassing the alpha units. The ultimate step of carrying out the entire enrichment in K-25, and dispensing with S-50 and Y-12 altogether, could not be taken because the final separation stages of K-25 had been countermanded when the K-25/Y-12 tandem scheme had been introduced; the maximum K-25 could achieve was about 20 per cent enrichment. After the war, however, the whole process was performed in a diffusion plant; this is simpler and less expensive than switching to an electromagnetic plant for the final stages.

Once they got going, the Oak Ridge plants processed some tens of tons of natural uranium to produce 60 kilograms of ^{235}U in a matter of weeks for the Hiroshima bomb.

9 Manufacture of Plutonium

Work on the plutonium alternative proceeded in parallel with that on uranium isotope separation, but almost entirely independently of it. This was because Groves imposed the principle of compartmentalization. Each person should work in his 'compartment' and his knowledge of other compartments should be restricted by his 'need to know'. This was good for security, but meant the loss of much fruitful discussion, and in practice there had to be compromise. Nevertheless the Met. Lab. in Chicago was confined to work related to the plutonium programme and was kept largely in ignorance of that on ^{235}U. The two programmes met at the top, at Bush's and Groves's level, and they also came together again later, at the stage of bomb design.

The work of Fermi and his colleagues on the uranium/graphite reactor provided the essential foundation, without which there could be no plutonium alternative. They had enough uranium and graphite of fair quality by July 1941 to start building a series of subcritical assemblies, along the lines of, for example, the earlier French work with uranium/water systems. The Americans called these 'intermediate experiments', and thirty were carried out in all, with improvements in design and in the quality of the materials, until the time was ripe to try to build a critical assembly, a self-sustaining reactor. The degree of success at any stage was indicated by the neutron multiplication factor, the k-value, as discussed in Chapter 4.

A lattice arrangement was generally used. Fermi started with a lattice constructed from graphite blocks and metal boxes filled with uranium oxide. The whole structure formed a cube with eight-foot sides, and contained about seven tons of uranium oxide. Although some neutron multiplication was observed, the k-value being 0.87, the result was not particularly encouraging. Fermi suspected impurities in the oxide, and chemical analysis proved him right. With purer uranium and other improvements, the measured k-values steadily rose. The introduction of a better grade of graphite in May 1942 gave $k = 0.995$, only just below the key figure of unity.

At last in July 1942 a value higher than unity, $k = 1.007$, was obtained. This meant that a critical assembly was possible in principle, though it would have to be very large, unless k could be increased still further.

The team moved to the Met. Lab. in Chicago in the summer of 1942 for the last lap, and joined another group under Samuel K. Allison, who were also making intermediate experiments.

Success was believed to be within reach, and to depend mainly on materials. Uranium metal, for example, should be superior to uranium oxide. Pure metal proved, however, to be difficult to prepare in quantity, and even though several manufacturers worked on it, it was not forthcoming until November. Even then there was only enough for the centre of the assembly, and oxide had to be used in the outer parts. Intermediate experiments gave $k = 1.07$ for a lattice of metal and graphite, and $k = 1.04$ and 1.03 for lattices of high purity oxide and graphite (two different grades), so the prognosis was excellent.

Construction of the critical assembly began as soon as the metal was available. The site was an old squash court under a football stadium in the middle of Chicago, and it was chosen in preference to a more isolated site being prepared in the Argonne Forest outside the city, so that the experiment could go ahead without delay. Fermi assured those concerned that the chain reaction would not get out of hand, basing his confidence on the knowledge that a small proportion of the neutrons do not appear immediately after fission, but are delayed, some by as much as a minute or more. Now if there are only just enough neutrons for the chains to be self-sustaining, then these delayed neutrons will be the limiting factor, determining how rapidly the chains branch and how fast the number of neutrons increases. The rate of build-up may then be as long as minutes or even hours, so there is ample time for action. A control rod, which absorbs neutrons, can be moved into the assembly, or in an emergency such a rod can be dropped into the device under gravity.

The story of Fermi's final experiment was reported in vivid detail four years later, and the following account picks out some of the highlights.

By 2 December everything was ready (Fig. 12). The assembly had been constructed and was only held in check by the control rod, a stick of wood wound with cadmium foil. At 10.37 a.m. Fermi ordered the first retraction of the rod by a short distance. The measuring instruments registered a rise in the number of neutrons,

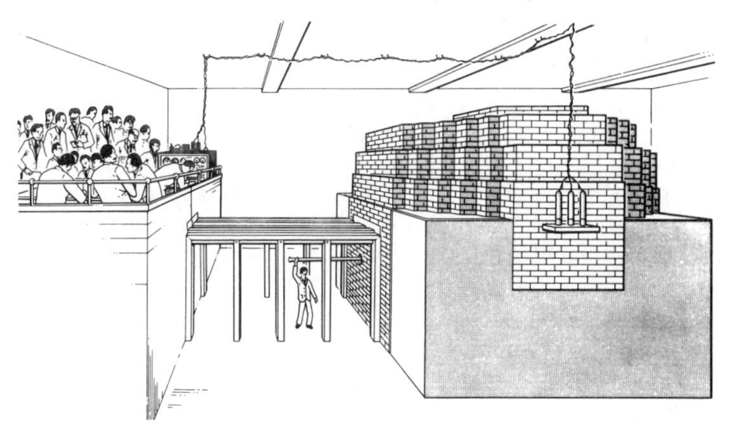

Fig. 12. CP-1, the first man-made nuclear reactor.

The main structure consisted of a lattice of uranium and graphite. Fermi ran the experiment from the balcony. The man on the floor pulled out a cadmium control rod step by step. Men on top of the reactor were ready to pour a cadmium solution on to it, if it got out of hand.

but then this levelled off. Fermi made it clear that this was what he expected. Step by step throughout the day the control rod was withdrawn, and each time the neutron count rate rose and levelled off. Finally at 3.25 p.m. Fermi said, 'Pull it out another foot. This is going to do it. Now it will become self-sustaining. The trace will climb and continue to climb. It will not level off.' For the next twenty-eight minutes the assembled scientists and engineers watched the count rate increase and go on increasing. By 3.53 p.m. the evidence was conclusive, and Fermi ordered the experiment to be shut down. Those present drank to success – in silence, but wondering if the Germans had got there first.

The demonstration had been exceedingly impressive. The engineers watching were convinced that here was no crackpot scientific fantasy, but a device capable of precise control, which they could take over and develop. Fermi's confident bearing, despite his scientifically cautious temperament, added to the impression.

In his book, *Atomic Quest*, Compton has recorded in rather flowery language the reactions of some of those present. Of Crawford H. Greenewalt, aged 40 and destined within a few years to become President of the great du Pont firm, he says: 'His eyes were aglow. He had seen a miracle.' In contrast to this was Volney

C. Wilson, a young, able, thoughtful, idealistic physicist. Of him, Compton writes:

> He was among those who had sincerely hoped that, even at the last moment, something might arise which would make it impossible to effect the chain reaction. The destruction it implied was a nightmare with which he was finding it hard to live . . . But Volney was a good soldier. He knew that if atomic weapons could be made we must be sure to have them first . . . but his face mirrored his inner conflict.

Of himself, Compton records no such conflict, no questioning of the objective of the Manhattan Project.

Later that day Compton phoned Conant. To pass the news without breaching secrecy he said in now-famous words, 'Jim, you'll be interested to know that the Italian navigator has just landed in the new world.'

'Is that so?' was Conant's excited response. 'Were the natives friendly?'

'Everyone landed safe and happy,' replied Compton.

There is a plaque at the University of Chicago to commemorate the success of the work. It says simply:

ON DECEMBER 2, 1942
MAN ACHIEVED HERE
THE FIRST SELF-SUSTAINING CHAIN REACTION
AND THEREBY INITIATED THE
CONTROLLED RELEASE OF NUCLEAR ENERGY

The reactor they had built, called CP-1, was a marvellous new tool for Fermi. Nobody had ever had one before. All kinds of interesting experiments could be carried out very quickly. It was like the heady days in Rome in 1934 when slow neutrons were first discovered.

There was a lot to be learned about the reactor itself, lessons which were to be invaluable for the large plutonium production reactors that were in the offing. For instance it was shown that CP-1 had an automatic safety feature: if it heated up it tended to shut itself off. It could also be used to test the suitability of prospective reactor materials, including samples of uranium and graphite, by the simple expedient of pushing them into the reactor and measuring their effect on the neutron level; this was much quicker than the methods available earlier.

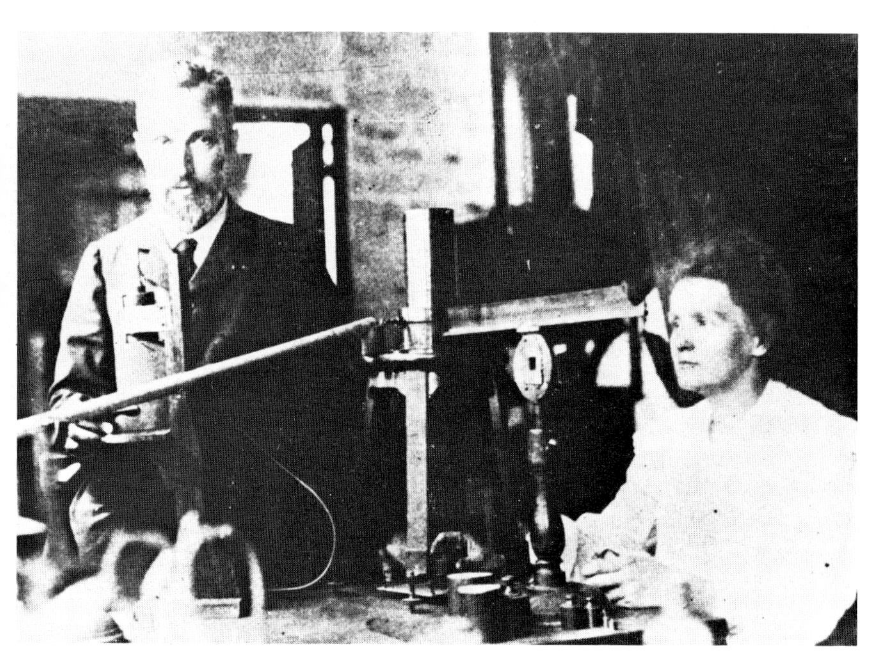

Plate 1. Pierre and Marie Curie, early 1900s.

Plate 2. Lise Meitner and Otto Hahn, Berlin, 1913.

Plate 3. Lord Rutherford, sketched by Otto Frisch.

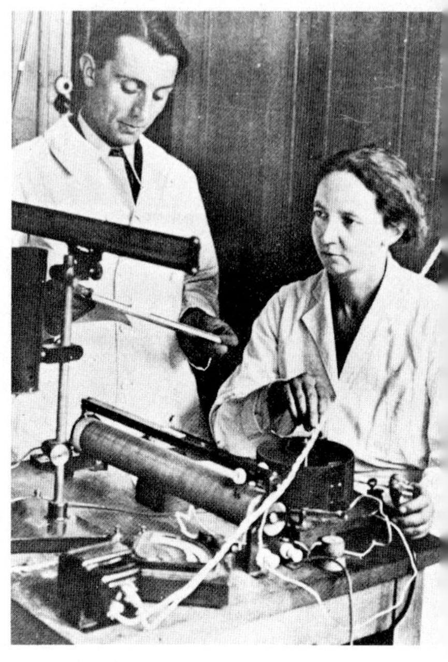

Plate 5. Frédéric and Irène Joliot-Curie, Paris, *c.*1935.

Plate 4. Rutherford's apparatus for the first artificial nuclear transmutation in 1919 (see Chapter 1).

Plate 6. The first nuclear accelerator ('atom smasher'), built by John Cockcroft and Ernest Walton in 1932 (see Chapter 1).

Plates 7, 8, 9. Copenhagen, *c.*1936. Pl. 7 (*above left*): an occasion at Niels Bohr's Residence of Honour. Pl. 8 (*above right*): Werner Heisenberg and Niels Bohr. Pl. 9 (*below*): Otto Frisch in the Niels Bohr Institute.

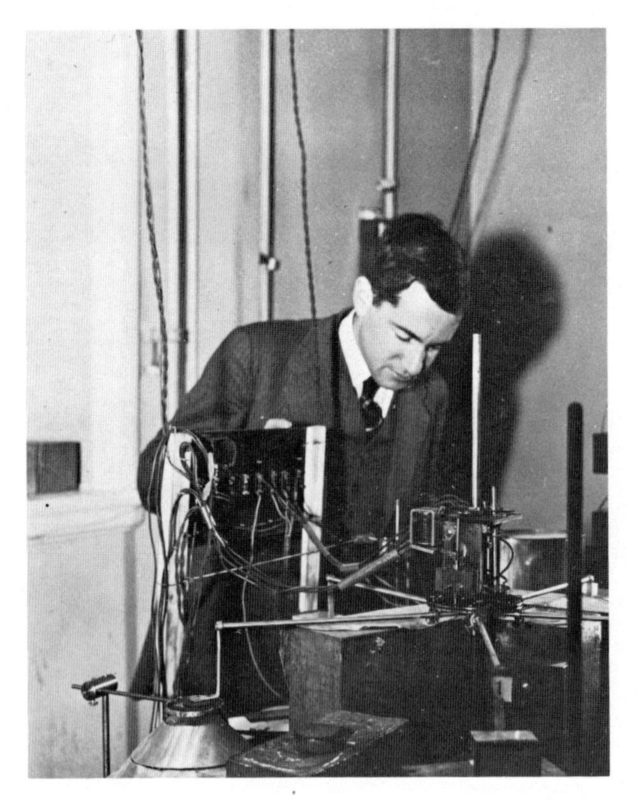

Plates 10, 11, 12. Leading Manhattan Project personalities. Pl. 10 (*above left*): Enrico Fermi. Pl. 11 (*above right*): Arthur Compton. Pl. 12 (*below*): Ernest Lawrence, Glenn Seaborg, and Robert Oppenheimer.

Plate 13. One of the alpha 'racetracks' in the electromagnetic separation plant for uranium isotopes (see Chapter 8).

Plate 14. James Chadwick and General Groves.

Plate 15. A small part of the Hanford site during construction in 1944.

Plate 16. John Cockcroft cutting the first sod at the Atomic Energy Research Establishment, Harwell, 1946.

Plate 17. Christopher Hinton at Dounreay, *c.*1957. In the background is the Dounreay fast reactor under construction.

After three months CP-1 was dismantled and re-erected as CP-2 in the new building provided for it outside Chicago. This was the beginning of the Argonne National Laboratory, which has contributed so much to nuclear science.

In anticipation of the success of CP-1, planning for plutonium production had started months earlier. As with the ^{235}U work, this leap-frogging method of progress was necessary if the end-product was to arrive in time.

The reactors for plutonium production had to be much more complicated than CP-1. For example, they would generate a million times as much heat, and would therefore require cooling. There must also be some means of withdrawing uranium when required, so that it could be sent to a chemical plant to extract its plutonium content. All this involved the introduction of structural materials and these would absorb neutrons; to compensate for this the reactor must be made larger. The Met. Lab. was the first in the world to face such problems.

By the late summer of 1942 there were three competitive reactor designs. The laboratory's Engineering Council had a design based on a helium coolant; Wigner proposed water cooling, while Szilard was a protagonist for liquid bismuth, an exotic choice which nevertheless had important advantages. The three coolants were very different, and this meant that many other features of the designs were different too.

Simultaneously with work on the reactor, Seaborg's team, also at the Met. Lab., were working on a manufacturing process to recover plutonium from the uranium in the reactor. Using their tiny traces of cyclotron-prepared plutonium they explored many possible separation methods, and eventually chose one devised by Stanley G. Thompson, one of Seaborg's Berkeley colleagues. (It depended on precipitation of bismuth phosphate.)

Up to this point nobody was better qualified to carry out the research and planning than the Met. Lab. itself, but the transition to the industrial scale was another matter. Conant complained to the Chicago scientists that they were going 'after elephants with a peashooter'.

It was to deal with the 'elephants' that the Manhattan Project had been created in the autumn of 1942. First Groves and the Army had been brought in, and then big industrial firms.

Du Pont went in under pressure from Groves, but with their eyes open, to take on plutonium manufacture. One of their vice-presidents, Charles Stine, told his senior staff that plutonium was essential to the nation's safety, and then went on to say, 'Du Pont is the only company that can do the task. We must do it, even though it may break the company.'

While the Met. Lab. accepted Groves and the Army with little demur – after all, one must expect to have the military around in wartime – some of them deeply resented du Pont. The refugee scientists' suspicions of big industry had already contributed to one outburst at the laboratory over the firm of Stone and Webster. Now there was the added fear that they were to be robbed of an exciting future at the heart of the plutonium programme.

For Compton, on the other hand, who carried the responsibility for the success of the work, 'it was a joy to be working now with the nation's first team'.

Du Pont had first to build pilot plants, comprising a reactor and a chemical separation plant, and then production plants. The Oak Ridge site in Tennessee was available for the former, but a larger site in a sparsely populated area was felt to be necessary for the latter. A search in December 1942 led to Hanford on a bend in the Columbia River in Washington State in the north-west, where 600 square miles of semi-desert were appropriated.

One of the first decisions du Pont had to make was between the rival reactor designs, and Szilard especially criticized them for taking so long to make up their minds. Yet it was a crucial decision, and a mistake might lose a lot of time. Wigner's water-cooled scheme finally won the day in early 1943. The design consisted of a block of graphite with aluminium tubes running horizontally through it. Filling most of the space in the tubes would be uranium metal 'slugs' in tightly fitting aluminium cans, and through the narrow gaps between the slugs and the inner wall of the tubes would flow the cooling water. In due course the reactor would be shut down and the canned uranium would be pushed out of the far ends of the tubes, where it would fall into a deep tank of water to await chemical treatment. The water in the tank would shield the operators from the radiation from the slugs.

While the basic idea for this reactor had originated in the Met. Lab., the detailed engineering was carried out at du Pont headquarters in Wilmington, Delaware. The main action rapidly shifted to Wilmington and the plant sites, Oak Ridge and Hanford. 'Whether the Chicago scientists liked it or not, the Metallurgical Laboratory had become a vital, but distinctly subordinate affiliate of the du Pont organization.' Wigner felt so strongly that he even tendered his resignation, but Compton persuaded him instead to take a month's holiday.

At this difficult juncture there was a revival of interest in heavy water. A decision had been taken in 1942, in co-operation with the Canadian Government, to manufacture it at Trail in British Columbia. This was partly at Urey's instigation in case graphite should prove unsatisfactory as a moderator. Du Pont were also favourably impressed with the prospects for a heavy water reactor and Groves had given them authority to build three plants for heavy water in the US.

Compton was therefore able to offer his discontented physicists a new outlet for their energies: the design of a heavy water reactor. Fifteen kilograms of the valuable material reached Fermi from Trail in the late spring of 1943, and when he tested it he was elated to discover that it absorbed almost no neutrons. Enthusiasm grew rapidly, and Urey, being at this point despondent about gaseous diffusion for uranium isotope separation, saw a heavy water reactor as the only hope for a bomb.

The scientists felt that du Pont meanwhile were floundering, lost in their own red tape. They could not understand why design of the water-cooled graphite reactor should take so long. The du Pont plans seemed to them to be over-engineered, too conservative, when the urgent need was for plutonium as soon as possible.

The situation threatened to blow up more violently than before. To meet the danger, Groves asked for a review by a special committee, which took place in August 1943. Grievances were aired, and a limited heavy water programme was approved, including the building of an experimental heavy water reactor to be known as CP-3 at the Met. Lab.'s Argonne site. But the main lines of the atom bomb programme remained unchanged, and did not include a route based on a heavy water reactor. Apart from anything else, the time-scale for producing large quantities of heavy water was too long.

While all this was going on, the chemists were making good progress in calmer seas. Seaborg and his men did not have the European refugees' prejudices against big firms, and got on well with their du Pont opposite numbers from the start. Moreover, their respective roles were clear and self-evident, and there was much exciting new chemistry for the Met. Lab. to do. It seems to be true, too, that chemists are temperamentally more pragmatic than physicists, less prone to perceive deep cleavages of principle.

Du Pont built the pilot plants at Oak Ridge during 1943. These were as small and simple as possible to prove the plutonium production process and make useful quantities of the new element. The reactor was designed to develop one megawatt of heat (equivalent to a thousand small domestic heaters), which implies the production of about one gram of plutonium a day. Improvements later raised these quantities several-fold.

Since a megawatt spread throughout the whole structure was still a comparatively low heat rating, air instead of water could be used as a coolant. Problems of corrosion and heat transfer, which were to be of great concern in the main water-cooled reactors at Hanford, were relatively simple to deal with, though the canning of the uranium slugs in aluminium for protection proved tricky.

The Oak Ridge reactor began operating on 4 November 1943 and was successful and trouble-free from the start. After a few weeks' running the first uranium slugs were pushed out and transferred to the chemical plant on 20 December.

This plant was different from any ever built before because it had to handle unprecedented quantities of radioactivity. Thick shielding was necessary, and all the chemistry had to be done on the far side of this. The general layout consisted of a 'canyon' containing a row of six heavy concrete cells, two-thirds buried underground. In each cell a process operation was carried out, and then the material was transferred to the next cell. Everything had to be done by remote control. Finally plutonium was obtained sufficiently pure to be handled in an ordinary laboratory, though still with great precautions against ingestion into the body. Despite the extreme novelty of the technology, and the fact that the chemistry had to be worked out with only a microscopic quantity of plutonium, the chemical plant was as great a success as the air-cooled reactor. By March 1944 it had produced several grams of plutonium.

The Oak Ridge installations provided information and experience for the full-scale plants at Hanford. Here it was originally planned to build six reactors and eight chemical separation plants, but the numbers were later scaled down to three of each. The reactors were designed to run at a power level of two hundred megawatts, compared with one megawatt at Oak Ridge, and required water cooling. There was also a plant at Hanford for manufacturing uranium slugs. For safety's sake, the different plants were all several miles apart.

Site preparation and construction began in the summer of 1943 and continued throughout 1944 and into 1945. Under the efficient Army and du Pont organizations the work rolled inexorably ahead. The operation was a huge one; at one stage there were no fewer than fifty-five thousand people living in barracks and trailers on the site. Plate 15 shows a small part of the site during the construction period.

There were still many technical problems requiring help from the Met. Lab., not the least being the canning of the uranium slugs. This was a tougher problem than at Oak Ridge, because the can had to withstand water instead of air, and to permit the transfer of much more heat to the coolant. There was fear, too, that a single can failure might put a whole reactor out of action. Henry S. Smyth recounts how on his periodic visits to Chicago he 'could roughly estimate the state of the canning problem by the atmosphere of gloom or joy to be found around the laboratory'. The problem was in fact solved just in time for the first Hanford reactor. On 13 September 1944 the constructors moved out and the operators moved in to begin loading the slugs into the first of the reactors. Fermi was again in charge, as he had been nearly two years earlier in Chicago. For the first few days all went well; the behaviour of the reactor, as the slugs were inserted and the control rods pulled out, was remarkably close to what had been calculated. By 2 a.m. on 27 September the power level was the highest so far achieved anywhere.

An hour later, however, the power had declined slightly for no apparent reason, and the operators were worried. The decline continued all through the day and by 6.30 p.m. the reactor had shut itself down completely. It began to recover the next day, but when the power level was raised again, it shut itself down once more. No such effect had been observed at Chicago or Oak Ridge.

The answer to this alarming new phenomenon was found astonishingly rapidly. The physicists had already given thought to the effect of the numerous species formed in the fission process on the neutron economy of the device, and it was appreciated that if any of them was a strong neutron absorber it would interfere with the neutron chain reaction. Now they analysed the behaviour of the Hanford reactor with this idea in mind. The key point was the recovery of the reactor, which indicated that, if a neutron absorber was responsible, it disappeared again in a day or so. A search through the information on the radioactive products of fission led to a species (a xenon isotope of mass 135) whose rate of radioactive decay (half-life, about 9½ hours) exactly matched the rate of recovery of the reactor.

It had taken only two days to reach this conclusion. Greenewalt of du Pont telephoned the news from Hanford to the Met. Lab. just as the scientists were clearing up to go home. Walter H. Zinn, who was in charge at the Argonne, held his men back to try to reproduce the Hanford phenomenon on their own CP-3 reactor. By running it at full power for long periods they were able to confirm Hanford's observations. CP-3, created partly at least as a result of the Chicago physicists' disputes the previous year, thus came to unexpected aid of the main programme.

While diagnosis was due to the scientists, cure depended on a decision taken by the du Pont engineers in the teeth of the scientists' opposition. With caution born of industrial experience, they had insisted on making the Hanford reactors much larger than they could logically justify, 'just in case'. There was room in them for a great many more uranium slugs, and this was just what was needed to overcome the effect of the xenon 'poison'. So the crisis subsided as quickly as it had arisen.

Nevertheless, there was delay while the effects of the new phenomenon were studied and mastered. Running the reactor up to full power proved a more complicated business than anyone had bargained for, and it took until the turn of the year. By this time the second reactor was also running, and the third was not far behind.

The first of the main chemical plants for processing the irradiated slugs from the reactors was just about ready when the first slugs were discharged. It was patterned on the pilot plant at Oak Ridge, but was far more massive. From the outside it looked like an enormous 800-foot long bar of concrete. Inside, it contained a row of forty tall concrete cells, with operating galleries running along the entire length of the building. The plant itself was a maze of tanks, centrifuges, pipes, and so forth, made of a special grade of stainless steel. As at Oak Ridge, all parts of the process had to be carried out remotely behind the concrete walls. Merely to train the operators for such an unusual plant was a colossal task in itself.

After extraction, the plutonium was passed to two smaller buildings for final purification and conversion to a suitable form (solid plutonium nitrate) for transport. Before the end of January 1945 the new element began to flow from the process buildings in ever increasing quantities. A number of kilograms had been produced by the summer of 1945, enough to be used in a test explosion and in the Nagasaki bomb.

Du Pont had taken on their assignment in the autumn of 1942. The Oak Ridge pilot plant built by them was functioning and producing plutonium just over a year later, and the large Hanford plant just over a year later still. This was amazing speed by any standards. Considering the newness of the technology, the scale of operations, and wartime shortages, it was fantastic.

10 The Weapons

How much ^{235}U or plutonium do you need for a bomb? What damage will it do? These were vital, practical questions, but the answers available at the start of the Manhattan Project were vague, and remained so for a long time.

The British had given some preliminary estimates for ^{235}U in the MAUD report on atom bombs: 10 kilograms of ^{235}U, of which 2 per cent might actually explode, releasing the same energy as 3,600 tons of TNT (though only doing about half as much damage).

To firm up the estimates, nuclear data were required for fast neutron processes, as distinct from the better-known slow neutron processes, which Fermi was dealing with in his uranium/graphite reactor. Fast neutron measurements had been made in Cambridge and Liverpool under the MAUD programme in 1940–1, and in several American universities rather later, but there were many discrepancies, and it was not easy to assemble a reliable set of data. One of the difficulties was the lack of sizeable specimens of enriched uranium and plutonium, which did not become available until 1944 from the Oak Ridge plants.

It was even harder to assess the efficiency of the bomb. The MAUD report had assumed 2 per cent, but figures up to 10 per cent were bandied about in 1941–2. What had to be done was to work out exactly what happens in the fraction of a second in which the supercritical mass is brought together and explodes; in particular the behaviour of the neutrons must be known in detail. Reliance had to be placed almost entirely on theoretical calculations. There was no possibility of small-scale experiments, because nothing below the critical mass can be made to explode; a test explosion is necessarily full-scale.

In the US, responsibility for the actual weapons was initially allocated to Compton. Lawrence forcefully recommended a relatively young Berkeley theoretician, J. Robert Oppenheimer, for this side of the work, and in May 1942 Compton put him in charge of a small group at the Met. Lab. At the time the task was not regarded as a major one; the optimists spoke of three months' work for twenty physicists.

Oppenheimer also had a strong team back at Berkeley to work on the theory of nuclear explosions, among them yet another redoubtable Hungarian, Edward Teller. This team very soon produced a startling new idea. As nuclear physicists they were well aware that energy could in principle be released not only by the fission of very heavy nuclei, but also by the fusion of very light nuclei, especially those of the isotopes of hydrogen. They saw that if material containing the light nuclei could be heated to enormous temperatures, like those inside the sun, the fusion process might take place. They also saw that an atom bomb might be just the way to obtain these very high temperatures. This was the beginning of the idea of the hydrogen bomb, or 'super' as it came to be called, still more powerful than the atom bomb.

Work on the super had to wait, but of immediate concern was Teller's suggestion that the explosion of an atom bomb might set off the fusion process in the hydrogen contained in the water vapour in the atmosphere, or in the water in the oceans. The whole lot might explode with unimaginable violence destroying the world. Unless this possibility could be conclusively disproved, an atom bomb could never be used.

This made it necessary to include studies of nuclear fusion in the programme, paving the way later for development of the super. The main line during the war, however, was always the fission bomb, which was anyway a prerequisite for the super.

As Oppenheimer grappled with his assignment, he began to discover its complexity and difficulty. In that first summer of 1942 it proved a hectic task merely to co-ordinate the fast neutron and other work which was scattered widely across the US, and keep it on the right lines. The natural answer was a special laboratory for weapon development where, thought Oppenheimer, some thirty scientists, mostly physicists, could work closely together.

In early autumn he put the idea to Groves, to whom it appealed immediately. It meant that the most sensitive part of the whole Manhattan Project could be located at an isolated site, where security could be particularly tight. Moreover Groves perceived a more extensive role for the proposed establishment. Not only research, but the design and production of the bomb could be carried out there, as well as the organization of test explosions.

The site, Los Alamos, in the southern Rockies, was chosen in November 1942. It lies in a strange, lonely region of extinct volcanoes, known to and loved by Oppenheimer from his youth. Canyons running north and south divide the area into isolated strips of flat-topped high ground, called mesas. Cliffs, often sheer, make the mesas difficult of access except along a few well-defined routes. The nearest habitation of any size, Santa Fe, is some thirty miles away. To Groves, Los Alamos seemed ideal.

Who should be Director of the new laboratory? Oppenheimer, the relatively inexperienced theoretician? Or somebody more practical, of greater prestige, perhaps with a Nobel prize? Besides, Oppenheimer had supported leftist causes, and both his wife and his brother were once Communist Party members. Yet over a period of several weeks nobody could offer Groves a better choice, short of taking Compton or Lawrence away from their vital tasks. Oppenheimer was to succeed brilliantly, and in retrospect his contribution appears to have been indispensable. One of his earliest assistants, John H. Manley, wrote of the 'astonishingly rapid transformation of this theorist, Robert Oppenheimer, into a most effective leader and administrator'. To many 'Oppie' became a charismatic figure.

Yet there was a question mark over him. Security kept him under surveillance throughout 1943 whenever he left Los Alamos. On 12 June they saw him visit a former friend and known Communist, Jean Tatlock, and spend a night with her. When they interviewed him on 12 September about an earlier espionage attempt at Berkeley, they found him reluctant to give the full facts. All in all, the data in their dossier on Oppenheimer added up to a classic example of a security risk.

Despite this, the security-minded Groves all but commanded Oppenheimer's clearance, describing him as 'absolutely essential', and did not yield when further suspicions were adduced later. Groves was never in doubt about Oppenheimer's loyalty to his country. Indeed, though Groves may not have known it, Oppenheimer would certainly not have wanted to pass American secrets to the USSR, because conversations with Plączek and other physicist friends who had lived in that country had convinced him that it was a land of terror and suffering.

All through this period the Los Alamos laboratory was being built up. The main outlines of the initial programme were blocked out in the spring of 1943. Besides fast-neutron physics and study of the explosion, the methods of preparation and fabrication of the actual explosives, presumably in metallic form, and the design of the bomb itself had to be worked out. This meant a great deal of chemistry, metallurgy, and ordnance engineering, far more than the physicists originally visualized. It was especially necessary to study plutonium, its purification, its conversion to metal, its properties as a metal, and its machining. The problems were exacerbated by the exceptional toxicity of the new element; special and often tedious handling techniques had to be devised. Moreover both ^{235}U and plutonium were for a long time in very short supply, and must be painstakingly recovered for re-use when experiments were completed.

For reasons already mentioned, there was also a small continuing programme on the hydrogen bomb.

Laboratories, workshops, and offices were built with wartime haste on the Los Alamos mesa, while a self-contained community with houses, shops, cinemas, schools, and churches was established on an adjacent mesa. Barbed wire surrounded the whole site, and wire within wire fenced in the 'Technical Area'. Not only the scientists but also their families were severely restricted as regards travel, mail, and outside contacts.

Precisely because the site was so physically secure, Oppenheimer was able to argue for a wide measure of free discussion inside the wire, and this was agreed to by Groves, an important concession in view of the novelty of the problems to be tackled, though he later described it as a mistake. Groves also yielded over the issue of making the site a completely military one, with everyone in uniform, because on this basis it proved impossible to recruit some of the key scientists Oppenheimer wanted. One of the recruits, Robert F. Bacher, made his letter of acceptance simultaneously a letter of resignation, valid on the day the laboratory became a military installation.

Recruiting was in any case difficult. Oppenheimer had to winkle people out of important jobs without being able to tell them what Los Alamos was going to do. The numbers required grew continually, and by the end of the war there were six thousand people living on the site. They included some of the most powerful mathematical and physical brains in the world. As at other nuclear sites, several of the key men were exiles from Europe, for instance Hans A. Bethe from Germany, who headed the important Theoretical Division, and Teller, who was to become 'the father of the H-bomb'. A visiting scientist spoke of 'the unique intellectual atmosphere' of Los Alamos.

After the 1943 Quebec Agreement, there was a small influx of British scientists, all top-flight men, including Peierls and Frisch.

Groves accepted them in his holy-of-holies because they could speed up the work, and they made a contribution out of all proportion to their numbers. But among them was Klaus Fuchs, a German refugee. He and an American, David Greenglass, were to betray the secrets of Los Alamos to the USSR. Fuchs had been drawn into Communism when a student in Germany in the early 1930s, largely as a result of his family's terrible sufferings under the Nazis. He was a very able mathematical physicist, and had joined Peierls in Birmingham in 1941, where under the cover of an apparently quiet, blameless life, he began passing information to the USSR a year later.

When he was sent to Los Alamos, Fuchs for some reason chose to let his link with the Soviet spy network lapse. Their contact man, Harry Gold, a naturalized American, made frantic efforts to locate him, and finally caught up with him early in 1945, when Fuchs visited his sister in Cambridge, Massachusetts. Fuchs then prepared a comprehensive report on the work of Los Alamos for Gold, and later handed over documents about the plutonium bomb and its impending test in New Mexico.

Groves has written, 'Since the disclosure of Fuchs's record, I have never believed the British made any investigation at all', but this accusation was quite unjustified. The British Security Services at first gave Fuchs only a qualified clearance, because they had been told of his Communist background; but police inquiries had turned up nothing suspicious about him since coming to Britain, and he was soon given a full clearance and then naturalized because his abilities were greatly needed.

As in Oppenheimer's case, the national interest as perceived at the time had been allowed to overrule Security's fears. What the authorities did not know was that the passion that had made Fuchs a Communist still burned and was to have its logical outcome in his telling the Russians all he knew about the nuclear work. Unlike Oppenheimer, Fuchs was not yet disenchanted with the USSR; that did not come until after the war, when it was to contribute to his detection and arrest, as is described in a later chapter.

The British team also included Niels Bohr, hiding his identity from the outside world under the pseudonym Nicholas Baker. He paid extended visits to Los Alamos after escaping from Denmark in a fishing-boat in 1943. He was one of many who had been warned by a courageous official at the German Embassy in Copenhagen, Ferdinand Duckwitz, of an impending round-up of Jews.

Everyone at Los Alamos was eager for his advice. The scientists there were acutely aware of building on theory, with no practical test explosions to check their conclusions. Bohr, with his deep insight into physics, was invited to go through their reasoning, looking for errors or oversights.

Almost alone at that time he began to look beyond the war to the problems the atom bomb would pose in the years to come. Once when asked about the Manhattan Project's chances he replied, 'Of course it will succeed. But what then?'

He saw hope in the very fact that full-scale nuclear war would be almost inconceivably destructive. He believed that this could be turned to account in inaugurating a new era of greater openness between the nations. Survival might depend on it.

Bohr discussed his ideas with Roosevelt, Bush, Lord Halifax, Lord Cherwell, and others, most of whom were sympathetic. To Churchill, however, Bohr's views were woolly idealism, especially his proposal that America and Britain should take a first step in openness by informing the Russians before any nuclear explosion took place. He was with difficulty dissuaded from treating Bohr as a gross security risk.

While Bohr's mind was ranging ahead to the political issues, Los Alamos still had to solve its technical problems. Sometimes these even threatened to kill the whole project, turning Oak Ridge and Hanford into massive white elephants. Suppose, for instance, there was a slight delay between the capture of a primary neutron and release of the secondary fission neutrons. Then there would be a similar delay between the successive steps of the chain reaction, and the whole process would develop too slowly to give an effective explosion, even with fast neutrons. (With slow neutrons the chains develop too slowly anyway, because it takes time for a neutron to travel from one uranium atom to the next, as was discussed earlier.) This particular ghost was laid to rest by cyclotron experiments in November 1943, which showed that the time interval was sufficiently brief not to matter.

Further doubts centred on the issue of pre-detonation. As had been foreseen in the MAUD report, pre-detonation of a bomb, leading to a 'fizzle' rather than an explosion, could result from the presence of too many stray neutrons, or too slow an assembly of the supercritical mass, or a combination of both.

One source of stray neutrons might be chemical impurities. These do not emit neutrons on their own, but some of them do so when mixed with uranium or plutonium. The tolerable quantities of such impurities are very small, so nuclear explosives, especially plutonium, must be strictly purified, and Los Alamos found itself in the chemical business of carrying out the processes required.

No amount of purification, however, can eliminate the neutrons that originate from the uranium or plutonium itself; as the MAUD report had pointed out, such neutron production will occur through the process of spontaneous fission (Chapter 6). Here there were some surprises. Thus Segrè found more fissions in ^{235}U, and hence more neutrons, at Los Alamos than at Berkeley. This was readily explained as a cosmic ray effect, the intensity of the cosmic rays being greater at Los Alamos, which lies at a height of 7,300 feet. Moreover if some of the fissions were due to cosmic rays there must be less spontaneous fission than had been thought, and this was encouraging.

There was a less pleasant surprise in the case of plutonium. When Segrè came to test material from the experimental Oak Ridge reactor, he found an unexpectedly high rate of spontaneous fission. Seaborg had already warned that this might happen. He had foreseen that, in addition to the ^{239}Pu isotope so far investigated, reactor plutonium would contain another isotope, ^{240}Pu, and that this might well undergo spontaneous fission relatively readily. Segrè's result was in line with this speculation.

The worry was that the proportion of ^{240}Pu would be even higher in the plutonium produced at Hanford for bomb manufacture. The uranium slugs were irradiated considerably more strongly at Hanford than at Oak Ridge, so as to produce more plutonium, but this meant that the plutonium itself captured more neutrons, thereby converting a greater proportion of the ^{239}Pu into ^{240}Pu. Calculations indicated that neutron emission by spontaneous fission of Hanford plutonium would exceed the Theoretical Division's specification for plutonium by a factor of several hundred.

This would make it impossible to explode Hanford plutonium by the simple gun method, originally suggested in the MAUD report, owing to pre-detonation. A totally different type of bomb would be needed.

An idea for this had fortunately already been suggested by Seth H. Neddermeyer in April 1943. Essentially it consists of surrounding a hollow sphere of ^{235}U or plutonium with high explosive. When this is detonated it thrusts all the nuclear material inwards towards the centre of the sphere, where it makes a supercritical, explosive mass (Fig. 13). The process was called

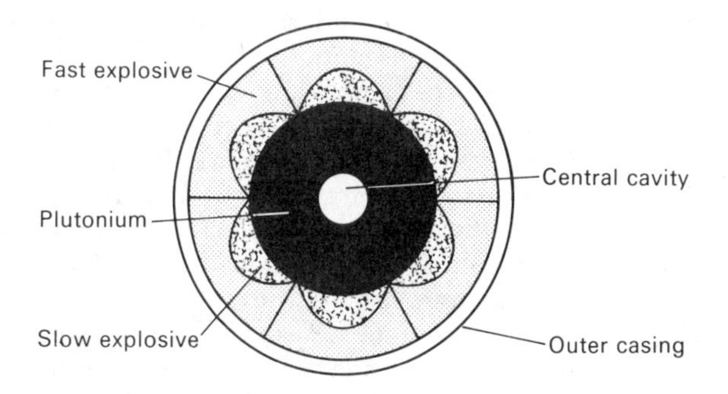

Fig. 13. The implosion concept.

The diagram shows a cross-section through an implosion bomb. The detonation of the ordinary explosives drives the plutonium together in the centre of the device, where it explodes. The combination of slow and fast explosives helps to ensure that all the plutonium moves at the same time and the same speed, so that it forms the necessary compact ball.

'implosion'. It had the potential to achieve what was wanted, but in 1943 it looked very difficult to bring off, because the implosion must be exactly symmetrical so that all the nuclear explosive moves inwards at the same instant. Otherwise it will not come together into a compact ball, which is the required supercritical shape. The gun method, based on familiar artillery concepts, seemed much more straightforward, though it needed a rather prodigal amount of nuclear explosive.

For some months only Neddermeyer's conviction kept the implosion method alive, and even he became discouraged. Then in the autumn of 1943 Teller pointed out that implosion would be still more economical of material than had been thought. This is because the enormous pressure generated compresses the nuclear explosive, making it more dense, so that less is required for criticality. The time taken for Hanford to deliver enough material for the first bomb might be reduced by several months. It was decided therefore to mount a major effort on implosion, and to bring in an expert on conventional explosives, George B. Kistiakowsky, to lead the work. In this area the British scientists at Los Alamos were able to make some notable contributions.

Even so progress was slow. Kistiakowsky, scheduling the work for the months ahead, made an imaginary final entry for late 1944: 'The test of the gadget failed. Project staff resumes frantic work. Kistiakowsky goes nuts and is locked up.' Actually late 1944 saw, not failure, but the first hopeful test result of the implosion process. The tests, of course, used inert material, not plutonium.

Meanwhile there was another doubt. After all the efforts to eliminate pre-detonation by stray neutrons, might not the opposite danger arise, and the stray neutrons prove too few in number to trigger the explosion? In implosion the nuclear material only finds itself for the briefest of instants in the ideal position to develop the chain reaction: would this give sufficient time for neutron multiplication, starting from just a few strays?

The obvious answer was to provide a sudden burst of neutrons at precisely the correct moment, and devices to achieve this were called 'initiators'. They contain two materials (beryllium and polonium in the 1945 device) which produce neutrons in profusion when intimately mixed, and the implosion process itself is used to bring about the mixing. Whether the initiator would actually work would, however, not be known until it was tried out in a nuclear weapon test.

With the uncertainties about implosion, Conant and Groves decided at the end of 1944 to allocate ^{235}U production to a gun-type weapon. This was regarded as a near-certainty, though wasteful of material. They foresaw having enough ^{235}U to promise a bomb by 1 August 1945, which they believed would release energy equivalent to ten thousand tons of TNT, but it might only be possible to make one more such bomb that year. This meant no test firing and no follow-up for some time, if the bomb was to be used that summer, a somewhat dismaying prospect.

As regards implosion weapons there would probably be enough plutonium to produce several during the latter part of 1945, if only the method could be made to work. The yield was pessimistically estimated at five hundred tons of TNT for the first implosion bomb, although the theoreticians hoped for ten times as much if all the bomb components functioned properly. A test firing was considered necessary to make sure the bomb actually exploded, and Independence Day, 4 July, was the date originally chosen for this. The test was code-named 'Trinity'.

The fixing of deadlines concentrated Los Alamos minds greatly. Choices had to be made, designs frozen, hardware constructed. After all the problems and disappointments of the previous months, everything now began to move forward. Supplies of ^{235}U from Oak Ridge and plutonium from Hanford were forthcoming in increasing quantities, and successfully converted into bomb material.

Great progress was made with implosion. The bombs themselves were built: 'Little Boy', the relatively slim gun-type ^{235}U bomb, and 'Fat Man', the implosion-type plutonium bomb (Fig. 14).

The Trinity test of a plutonium device was carried out in the Alamogordo desert in New Mexico at half-past five in the morning of 16 July 1945, after several postponements – nine days, three days, an hour, thirty minutes. For the Los Alamos men, and above all for Oppenheimer, this was the crunch. Had they got it right? Had they overlooked anything? Would the work of tens of thousands of men and women and the expenditure of thousands of millions of dollars end in a flop or a triumph? The tension was enormous as zero hour approached. Groves, to calm Oppenheimer, kept taking him outside to check on the weather, which was indeed a source of concern. Fermi, by contrast, showed no sign of strain. Ever the scientist, he said that Trinity was a worthwhile experiment whatever the result; if the bomb failed to go off it would show the impossibility of a nuclear explosion.

General Thomas F. Farrell, Groves's deputy, was in the control shelter ten thousand yards from the explosion with some of the key men. In his report he said:

> Everyone in that room knew the awful potentialities of the thing they thought was about to happen . . . We were reaching into the

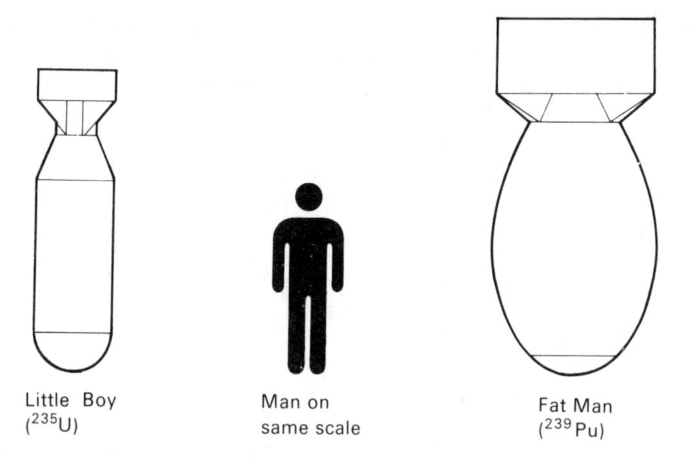

Fig. 14. The Hiroshima (left) and Nagasaki (right) bombs.
The all-up weights were 4 and 4½ tonnes respectively. The nuclear explosive constituted only a small fraction of this.

> unknown and we did not know what might come of it. It can safely be said that most of those present – Christian, Jew and atheist – were praying and praying harder than they had ever prayed before.

Trinity was successful beyond all expectations. The observers, the first to see the now familiar fireball and mushroom cloud, were overawed. To quote again from Farrell's report:

> The effects could well be called unprecedented, magnificent, beautiful, stupendous and terrifying. No man-made phenomenon of such tremendous power had ever occurred before. The lighting effects beggared description. The whole country was lighted by a searing light with the intensity many times that of the midday sun. It was golden, purple, violet, gray and blue . . . It was that beauty the great poets dream about.

Surprised by the brilliant light and absorbed in watching it, some were taken unawares and knocked flat by the shock wave thirty seconds or more later. This was followed by an 'awesome roar' which reverberated through the hills for several minutes. In cold scientific fact the energy released was several times as great as the upper limit of five thousand tons of TNT estimated by the theoreticians.

After the explosion, Bush, Conant, and Groves silently shook hands. Kistiakowsky released his pent-up emotions by embracing Oppenheimer with shouts of glee. The general mood has been described as one of 'elation tempered with solemnity'. Farrell spoke of a feeling among those present that they 'should dedicate their lives to the mission that [the new force] would always be used for good and never for evil'.

Farrell's first words to Groves after the explosion were, 'The war is over.'

'Yes, after we drop two bombs on Japan,' replied Groves.

11 The Other Side of the Fence

Throughout the war, Heisenberg and his colleagues in Germany, the Uranverein, worked under the comfortable illusion that they were the world leaders in the race to demonstrate a nuclear chain reaction. They were unaware of the scale and speed of the Manhattan Project.

Early in the war they had a lead of nearly two years over the Americans. By mid-1940 their perception of the main lines of attack was similar to that of the S-1 Committee in the first half of 1942. They had several ideas for uranium isotope separation, and they envisaged a uranium/heavy water reactor which could produce a new element capable of fission. They mistakenly discarded the idea of a uranium/graphite reactor, but even so they had several possible routes to nuclear explosives.

Since, however, many of them, including Heisenberg, were more interested in science than in making an atom bomb for Hitler, the main objective of the Uranverein became the building of a critical uranium/heavy water assembly. This was exciting scientifically and of immense potential post-war interest, but seemed unlikely to lead to a weapon before the war ended.

Uranium isotope separation in Germany similarly became directed more towards reactors than explosives. Uranium enriched in the ^{235}U isotope was seen not so much as bomb material as an answer to the shortage of heavy water, because it could be used in combination with ordinary water instead of heavy water; it is indeed so used in the majority of present-day power reactors.

The German approach to enrichment of uranium in ^{235}U diverged in several ways from the British and the American. Harteck's group experimented at length with gaseous thermal diffusion before acknowledging failure, and then turned to the centrifuge method for the rest of the war, achieving a modest success. The British, on the other hand, abandoned both methods rather quickly, while the Americans found the former unpromising, and eventually rejected the latter as too difficult to engineer on the scale required for the Manhattan Project.

Of the methods actually used in the US, gaseous diffusion through a porous barrier was, oddly enough, scarcely considered by the Germans though they were well aware of the principle; electromagnetic separation was pursued under Post Office (!) auspices; and liquid thermal diffusion apparently did not occur to them.

The Uranverein also had a novel method, briefly considered but rejected by the British, which they called an 'isotope sluice'. This was invented by Bagge, who, unlike some of his colleagues, really hoped to use it to produce material for a bomb. It employs what is usually called the time-of-flight principle, which has been known to science since Fizeau used it in 1849 to measure the velocity of light. Essentially it involves two rotating shutters in the path of a beam – of light in Fizeau's case, but of uranium atoms in a high vacuum in Bagge's application. If each shutter opens momentarily, then the beam can pass right through the system provided the opening of the second shutter is timed to occur slightly later than that of the first; the delay must correspond exactly to the time-of-flight between the shutters. In Bagge's scheme the time-delay was adjusted to let only the fastest uranium atoms through, and these tended to be the lighter ^{235}U atoms rather than the heavier ^{238}U atoms. Bagge ultimately managed to separate a few grams of the uranium isotopes in this way.

The Germans considered at least three other methods besides those mentioned, but none came to anything. They achieved small-scale separations by the centrifuge as early as August 1942 and the isotope sluice in July 1944, and but for Allied bombing might have gone on to scale them up. Bagge in particular had a frustrating time with air attacks; both his first two models were destroyed, and three times he had to move everything to a new location.

The status of the electromagnetic method in Germany was rather curious. It appears to have been largely ignored by the Uranverein, and their post-war surveys of German wartime work on isotope separation do not even mention it. Yet it was evidently heading for success by the end of the war.

The concept was basically the same as Lawrence's in the US, and it was put forward by Baron Manfred von Ardenne early in 1940, a year ahead of Lawrence. Ardenne, like Lawrence, was a technical go-getter. Wanting funds to exploit his idea, he approached the Minister of Posts, Wilhelm Ohnesorge, suggesting that he had a method of making a few kilograms of ^{235}U for an atom bomb.

Ohnesorge informed Hitler, but the time was unpropitious; in the summer of 1940 the war seemed as good as won for Germany, and there was little interest in new weapons. Hitler was sarcastic, and Ohnesorge came away with a flea in his ear, though undeterred.

Meanwhile Heisenberg and Weizsäcker were getting anxious about Ardenne's activities. The last thing they wanted was for Hitler to call for a crash atom-bomb programme. Weizsäcker, presumably with his tongue in his cheek, accordingly approached Ardenne with the message that for technical reasons a ^{235}U bomb did not seem possible. Ardenne bowed to what he believed to be superior knowledge and switched his efforts to nuclear accelerators. Nevertheless he returned to his electromagnetic separation idea later. In 1945 the Russians were sufficiently interested to round him up and take him to Russia, along with a number of other German nuclear scientists.

None of the German work on uranium isotope separation made enough progress to contribute to their reactor-building programme, so, as in the US but for a different reason, work on the latter proceeded independently. It included twenty-two experiments with uranium/moderator assemblies, listed by Heisenberg as follows:

- Three early attempts in 1940–1 to make a critical assembly, including Harteck's uranium oxide/dry ice experiment described in Chapter 5.
- Ten experiments under Heisenberg at the Kaiser Wilhelm Institute in Berlin. The first five were performed in 1940–2 in a laboratory specially built in 1940, and called the 'Virus House' to scare away unwanted visitors. When bombing became severe, a massive concrete bunker was built; this was available for the next five experiments in 1944.
- Four experiments under Heisenberg and L. R. Döpel at Leipzig University in 1941–2.
- Four experiments under Diebner, mainly at the Army Ordnance Laboratories at Gottow in 1941–3.
- A final experiment under Heisenberg in a cave in the village of Haigerloch in south Germany in March 1945, after his group had had to leave Berlin in face of the Russian advance.

There were thus three principal series of experiments in Berlin, Leipzig, and Gottow respectively. Those in Berlin had a stack of alternate layers of nuclear fuel and moderator. This is less efficient than a lattice, such as Fermi used, but confers something of the same benefit. Those in Leipzig also used layers, but fewer and arranged spherically, like the layers of an onion. This arrangement lends itself well to theoretical analysis of the results, though it is somewhat troublesome to construct, requiring a series of aluminium spheres to separate and support the layers of fuel and moderator. Only the Gottow experiments and that in Haigerloch used a lattice. The whole structure, of whatever type, was generally lowered into a tank of water or surrounded by paraffin wax; some of the neutrons which would otherwise have escaped were reflected back into the assembly by these substances.

As in the US, progress was conditional on the supply of materials. Initially, in all three German series, uranium oxide had to be used for lack of the metal, and paraffin wax for lack of heavy water. It was almost a foregone conclusion that there would be no measurable neutron multiplication, but some useful information could be obtained, and it was in any case wise to check whether this easy approach might not be successful.

Uranium metal was prepared by the firm of Degussa in Frankfurt, who were able to draw on earlier experience in preparing thorium metal. At first they supplied the uranium in powder form but later in plates and cubes. In all they provided fourteen tons of high quality material, ample eventually for the physicists' purposes. It is strange to reflect that if Fermi had had Degussa's uranium, it might have speeded up his work by several months.

For heavy water the Germans had to wait for fresh production from the Norwegian plant, because the French had taken the original Norwegian stockpile of the material. Rjukan, where the plant was situated, held out longer during the German invasion than any other town in southern Norway, and was not captured until 3 May 1940. The plant itself was intact, but had to be modified to increase heavy water production, and this took about a year to come into effect.

News of activity at Rjukan reached Britain in 1941, and served to spur on the Allied nuclear effort. It also led to a decision to destroy the German source of heavy water. First came British and Norwegian commando attacks, one of the war's great epics, and then American bombing, which finally closed down production in November 1943. By then the Germans had obtained a little over two and a half tons of heavy water, which left them rather short. To claim that the crippling of the Norwegian plant saved the Allies from nuclear attack would, however, be going much too far; it prevented the Germans making a nuclear reactor, but they had barely even thought about the further very big step from a reactor to a bomb.

Uranium metal and heavy water became available in Germany in sufficient quantities for incorporation in the physicists' assemblies in 1941. The first supplies of metal went to Berlin for uranium metal/paraffin wax experiments, and those of the heavy water to Leipzig for a uranium oxide/heavy water experiment. The latter gave an encouraging result; although no definite neutron multiplication was observed, it was possible to conclude that this was only because the aluminium in the structure absorbed too many neutrons.

Heisenberg was now convinced that a successful 'uranium burner' was just around the corner, and might be used to manufacture 'explosives for atom bombs'. He also had the idea that the scientists of the world could hinder such a development, as he and his colleagues were indeed doing in Germany.

He and Weizsäcker decided to approach Bohr in Copenhagen, apparently with the intention of suggesting that Bohr should use his influence to achieve a moratorium on atom-bomb work by scientists on both sides in the war. Bohr's son, Aage, however, categorically denies that the Germans made any such proposition. Certainly Heisenberg and Weizsäcker had to be very careful what they said to Bohr for security reasons, and they may not have been explicit enough. Moreover Bohr was so shocked and appalled to learn that an atom bomb seemed feasible, and that Hitler might get one, that he could think of little else throughout the interview. Nature, which he had delighted to study all through his life, seemed suddenly no longer benign, but full of menace.

One of Heisenberg's colleagues saw the visit in a different light, as due to Heisenberg's troubled conscience, saying that 'the high priest [of German physics] has gone to seek absolution from his pope'.

Whatever may have been in the minds of the protagonists, the interview was an abortive one, and the Germans returned home, Heisenberg to continue his attempts at reactor building.

This was a time of crisis for the Germans' war effort. Against Russia their blitzkrieg tactics had failed, and they faced a long drawn-out struggle, imposing severe strains on the economy. Could the uranium project be allowed to continue? Rather indecisively, control of the Uranverein was transferred from the Army to a civilian body, the Reich Research Council. Esau, of all people, who had been ousted from the uranium project in 1939, was put in charge. Diebner, however, was unaffected, since he was working in one of the Army's own laboratories.

There ensued a period of confusion and uncertainty. Nevertheless, Döpel and Heisenberg pressed on in Leipzig with their first uranium metal/heavy water assembly, and by May 1942 it was ready for testing. A total of 572 kilograms of uranium metal powder along with 140 kilograms of heavy water were introduced into their spherical aluminium structure, and the whole contraption was lowered into a pool of ordinary water. This time a positive result was obtained: there was a 13 per cent increase in the number of neutrons at the surface of the sphere, over those injected at the centre, corresponding to a neutron multiplication factor (k-value) of 1.01, just above the key value of unity. Döpel and Heisenberg calculated that they had only to build a larger pile to the same design, with roughly ten tons of uranium metal and five tons of heavy water, and they would have a nuclear reactor.

Three months later, Fermi in America was to report a k-value greater than one for a uranium oxide/graphite assembly. Unknown to one another the two projects were running neck and neck, but from then on the American programme streamed ahead.

The crucial German decisions on the future size of their programme, and even whether it should continue at all, were taken at a secret conference with Albert Speer, the Minister of Munitions, in Berlin on 6 June 1942, exactly a fortnight after the equally crucial meeting in Washington, when the decisions were taken to build industrial-scale plants in the US. Heisenberg had encouraging results to report, and was able to say in reply to a question from Field Marshal Erhard Milch that a bomb 'as big as a pineapple' would destroy a large city, but he also made it clear that the timescale for manufacture would be several years. Speer was sufficiently impressed to authorize funds for the Uranverein's work to continue on its existing scale, but in view of Hitler's orders to concentrate on tasks which would produce military results quickly, he did not offer massive support. That went instead to the flying bombs and rockets.

Meanwhile Döpel and Heisenberg's uranium/heavy water sphere had been left immersed in its water tank in Leipzig, where it came to a sticky end. After twenty days it began to produce a stream of hydrogen bubbles, evidently due to a water leak, allowing water to react with the uranium metal powder. The outer casing was therefore opened for inspection. This, however, allowed air to rush in, and the result was a regular firework display of burning uranium. This was doused with water, and the sphere was lowered into the water tank to cool it off. Instead, it got hotter. Watching it, the physicists saw it shudder and start swelling. They ran for their lives. A few seconds later the sphere exploded, showering the building with burning uranium and setting it alight. Not only did Döpel and Heisenberg lose most of their uranium and heavy water, but they had the galling experience of being congratulated by the Leipzig fire brigade on an amazing display of atomic fission, when it was in fact only the chemistry that had got out of hand.

That was the end of the work in Leipzig, and subsequent experiments were made in Berlin with greater regard to uranium chemistry.

Meanwhile Diebner, having been slowly squeezed out of the Kaiser Wilhelm Institute and feeling that the Uranverein physicists were too theoretical and too slow, was carrying out his rival series of experiments under Army auspices, without informing Heisenberg. Diebner, though not in the top flight, was a sound practical scientist, and his experiments were well performed. His most notable contribution was the use of a lattice. His first uranium oxide/paraffin wax assembly enabled him to demonstrate the superiority of the lattice arrangement, and his second experiment in the spring of 1943 with 108 uranium metal cubes embedded in heavy ice gave a k-value of 1.08. By using heavy ice, Diebner eliminated the need for supporting materials which would waste neutrons, and this was partly why he achieved a considerable improvement on Döpel and Heisenberg's final Leipzig k-value of 1.01.

For his next experiment late in 1943 Diebner had the simpler idea of hanging strings of several uranium metal cubes on wires in a tank of heavy water, and was able to obtain still better results. With more heavy water he might have gone on to make a critical assembly, but he was frustrated by the destruction of the Norwegian heavy water plant and had to turn his stocks over to Heisenberg and his colleague, Karl Wirtz, for their experiments in the bomb-proof bunker now ready in Berlin.

These now came under Walther Gerlach, a physicist with the confidence of the Uranverein who had replaced the unsatisfactory Esau. Gerlach made it his aim to maintain as much scientific research as possible, so that German science could rise again after the war.

Throughout 1944 Heisenberg and Wirtz still used their layer arrangement to facilitate comparisons with their earlier results, but perhaps also because Diebner's lattice was 'not invented here'. One innovation in their penultimate experiment was to have a neutron reflector partly of graphite; its success made them begin to wonder whether they had been right in rejecting graphite as a moderator. By the end of the year they were obtaining *k*-values of 1.08 and 1.09, and a nuclear reactor seemed within reach. They could still increase the amount of uranium, making in all one and a half tons each of uranium and heavy water, and there were other possible improvements including at long last the use of a lattice.

It was around this time that, as their armies advanced across France, the Allies obtained their first reliable picture of the state of progress of the German project. They had been nagged by the fear that the enemy might yet turn defeat into victory with an atom bomb. True, Allied Intelligence had no evidence of a large-scale enterprise, but that could be due to super-secrecy. Surely the Germans must be well along the road the Americans had taken, perhaps even in the lead!

To find out, the Americans sent a small mission into Europe to 'look into the atomic bomb development in Germany'. Its chief scientist was Dutch-born Samuel A. Goudsmit, one of the very few atomic physicists of standing in the US who had *not* been drawn into the Manhattan Project, and who could not therefore reveal any major secrets if captured. His parents had died in Nazi gas-chambers so he was strongly motivated.

The mission was named 'Alsos'. This means 'grove' in Greek, as someone happened to observe, giving General Groves a momentary qualm. The mission's mode of operation seemed very strange to the military. 'They could not understand how we could know, in advance, just which of the enemy scientists had the information we wanted,' wrote Goudsmit. 'To an outsider, a professor is a professor, but we knew that no one but Professor Heisenberg could be the brains of a German uranium project.' Peierls and his colleagues, it is interesting to note, had earlier given British Intelligence a list of sixteen scientists likely to take part in any German nuclear programme, and their only mistake was in failing to foresee that one of the sixteen would be excluded on racial grounds.

Alsos moved into Paris and Brussels immediately those cities were liberated, but it was not until Strasbourg was taken towards the end of November 1944 and they were able to go through the scientists' files at the University, including some notes of Weizsäcker's, that they found what they were after. Goudsmit records that as he and a colleague were reading them through by candlelight, they let out a yell at the same moment. 'We had both found papers that suddenly raised the curtain of secrecy for us. These were not secret documents; they were 'just the usual gossip between colleagues', but they showed unmistakably that the German project was on an academic scale, that there was no large effort, and that Germany was not likely to have an atom bomb for a long time.

While the Allied armies were held up in the winter of 1944/5, Heisenberg and Wirtz were poised for their final experiment. All was ready by the end of January 1945, but by then the Russian armies were approaching Berlin, and panic was rising in the capital. On 30 January, Gerlach gave orders to dismantle the uranium/heavy water assembly and take it south to Stadtilm in the middle of the country, where Diebner had established himself and was endeavouring to continue with his own experiments. Wirtz, however, was afraid that Diebner would try to take over the Kaiser Wilhelm Institute's materials. After much coming and going Wirtz finally got permission to take his convoy of lorries to the small village of Haigerloch, south of Stuttgart.

Here in a cave in the rock, Heisenberg and Wirtz, joined once more by Weizsäcker, put together their last assembly. They obtained their best result so far, 6.7 times as many neutrons coming out of their assembly as had been injected into it, and a k-value of 1.11. Clearly they were not far from criticality. Gerlach, informed in Berlin by telephone, said excitedly, 'The machine works.'

With some extra heavy water the physicists at Haigerloch still hoped to reach criticality before the end of the war, and they applied to Stadtilm for help, but Diebner had left for Hitler's 'Bavarian Redoubt' and his whereabouts was unknown. Soon after, the Allies overran the whole area, and rather pointlessly blew up the Haigerloch cave.

The persistence of the Uranverein even while their country was disintegrating around them is amazing. They singlemindedly pursued their scientific objective in the belief that they were ahead of the world, and that the Allies would be eager to learn their secrets.

The Alsos team, once they could get into Germany, visited every significant site, where they saw evidence of scientific work of the highest standard. Of the bunker laboratory in Berlin, Goudsmit wrote that, even stripped of its equipment, it 'gave an impression of high-grade achievement'. What they did not see was a large industrial programme.

They encountered very varied reactions from the German scientists. In Heidelberg Goudsmit was astonished to find how much pure physics his old friend Bothe had done in wartime, and they enjoyed talking it over together, but as soon as war research was broached Bothe said, 'We are still at war . . . If you were in my position you would not reveal secrets either . . . I have burned all secret documents.' Goudsmit did not at first believe that a scientist would destroy his experimental results, but concluded after a thorough investigation that Bothe had told the truth.

Bothe was a loyal German, but no Nazi, and had been dismissed from his professorship in favour of the top Nazi in Heidelberg, an insignificant scientist by the name of Wesch. Wesch's reactions when captured were in total contrast to Bothe's and typical of many Nazis'. He immediately offered his services to the Allies, wrote a lengthy and pompous report, and claimed not to have been a Nazi at heart.

Alsos's most important target was, of course, the Haigerloch area, but it lay in the zone allocated to the French. Groves, knowing that Joliot, France's foremost nuclear scientist, had become a Communist during the war, feared that the French might hand captured nuclear information, materials, and even scientists to the Russians. To forestall the French, therefore, a special Alsos unit made a dash for the area towards the end of April, took Hahn, Weizsäcker, Wirtz, and a few others prisoner, and removed the uranium, heavy water, and graphite, besides all the technical reports. A few days later Diebner, Gerlach, and Heisenberg were captured in Bavaria.

Heisenberg told Goudsmit, 'If American colleagues wish to learn about the uranium problem, I shall be glad to show them the results of our researches if they come to my laboratory.' The situation was deeply ironical, but the Germans could be given no hint of the true position, and Goudsmit fostered the idea that the Allies wanted their help in order to catch up.

Ten of the most significant German nuclear scientists were interned in England at Farm Hall, a country house in Huntingdonshire, and were there when the news of Hiroshima swept away their illusions.

12 Hiroshima and Nagasaki

In November 1944 the Alsos mission reports from Strasbourg had shown that there could be no German atom bomb, and any remaining doubts were dispelled early the following year by further Alsos reports from inside Germany. America was no longer in a race with the Nazis. 'Isn't it wonderful that the Germans have no atom bomb?' said Goudsmit, the leader of the mission, 'Now we won't have to use ours.'

'You don't know Groves,' replied one of its military members. 'If we have such a bomb, then we'll use it.'

Unknown to Alsos, the first bombs would probably be too late anyway for the war in Europe, which was to end on 8 May 1945. This meant that Japan, not Germany, was the prospective target.

In the eyes of many of the scientists, especially Compton's turbulent staff, this changed the moral complexion of the situation completely. Their principal motivation had been the fear of an enemy bomb. Germany might have made one, but not surely Japan. Was there any longer any justification for America to use it?

Perhaps some of them would have taken a different view if they had been aware that there was a Japanese atom-bomb project. This did not come to light until the 1970s, and our information about it is still sketchy. It is known that the Japanese nuclear physicists, led by Nishina, approached the Army in September 1940 and obtained funding for what was described as fairly large-scale nuclear research. It is also known that there was a major appraisal of bomb prospects in a report by a 'Physics Colloquium' in March 1943, which concluded that it would take Japan ten years to develop an atomic weapon and that even the US could not produce one in the current war. Despite this, Nishina's project continued all through the war, and the Navy funded a second project under Bunsabe Arakatsu at Kyoto University. There was also prospecting for uranium.

As regards the actual content of the Japanese programmes, it is known that Nishina's group built a small gaseous thermal diffusion plant to separate the uranium isotopes, and it was apparently hoped that the slightly enriched uranium product could be used in a uranium/light water reactor. Just when the plant was ready to operate, however, in April 1945, the building was wrecked in an air raid.

The work had clearly not got very far by the end of the war. Japanese science, strong in some fields, lacked a broad enough base, and Japanese industry was under too great a strain to be of much help. The scale of the nuclear work therefore remained small, and the American scientists were quite right to discount a Japanese atom bomb.

However, to the US Army men, including Groves, this was no reason for America to refrain from using her bomb. They had few scruples about exploiting it to the best military advantage. The war in the Pacific was tough. There had been 120,000 Japanese and 80,000 American casualties in the fighting to capture the island of Okinawa, and the invasion of the mainland of Japan was expected to cost a million American casualties alone. The sooner Los Alamos provided the means to shorten the war, the better.

The military tended to think of the atom bomb simply as an economical means of delivering the equivalent of twenty thousand tons of TNT. Some of the scientists, on the other hand, used the phrase 'an absolute weapon', meaning one which is absolutely decisive, rendering all other weapons futile. They saw the atom bomb as marking a watershed in history. They foresaw a new age, the atomic age, in which the energy of the atomic nucleus would bring great benefits to mankind. Was the world to be introduced to this new age through an atom bomb on Japan?

Each side had a case, and a collision was inevitable as the dates for the first weapons drew closer. Behind the immediate clash over the use of the bomb against Japan, lay the whole question of nuclear weapons and nuclear power in the post-war world.

The US Government did nothing very much about either issue until it was clear that the Manhattan Project was going to succeed, that is to say, until the spring of 1945, after Roosevelt's death. In early May, Bush, who had been trying for months to get something moving in this area, persuaded the new President, Harry S. Truman, to set up a so-called Interim Committee under the Secretary of War, Henry L. Stimson, to advise the Government. It was a high-level body, with three scientific members (Bush, Conant, and Karl T. Compton, who was Arthur Compton's brother and like him a Nobel laureate), and to assist them they had a Scientific Panel consisting of the top scientific brass of the

Manhattan Project, Arthur Compton, Fermi, Lawrence, and Oppenheimer.

The Interim Committee were busy men, but they found the time to consider carefully how the bomb should be used in the Pacific war. After examining the various alternatives, they recommended unanimously that it should be dropped on Japan without prior warning of the nature of the weapon. The aim was to achieve the maximum psychological impact in the hope of a Japanese surrender.

After the war, unworthy motives were to be imputed to the Interim Committee and others: their main concern was said to be to justify the cost of the bomb to Congress or to forestall a Russian declaration of war on Japan. Obviously thoughts like these must have passed through the minds of the men in question, but the record shows that the decision to drop the bombs was strategic, not political.

The members of the Scientific Panel reported back very guardedly to the staffs at their establishments, apparently without even telling them of the recommendation regarding the use of the bomb. Compton predictably ran into trouble at the Met. Lab. He invited his men to put their ideas on paper, and they fell to with a will. Unlike those involved in actual production at Oak Ridge, Hanford, and Los Alamos, they were under no special pressure, and they were able to drop their usual work for some weeks and go into committee.

One of their committees, which included the indefatigable Szilard, produced a particularly significant report, the Franck report, named after the refugee scientist who acted as its chairman, James Franck. This drew on ideas which had been brewing and maturing for at least a year. The main theme was the desirability of an 'international agreement on total prevention of nuclear warfare', to which considerations of the use of the bomb against Japan should be subordinated. 'The way in which the nuclear weapons now being secretly developed in this country are first revealed to the world', said the seven authors of the report, 'appears to be of great, perhaps fateful importance.'

They took the view that to use the bomb without warning against Japan might cause such international revulsion as to prejudice all hope of an agreement, and would not necessarily bring the war to an end. Instead they proposed a demonstration of the new weapon 'before the eyes of representatives of all the United Nations, on the desert or a barren island'. America could then say, 'You see what sort of a weapon we had but did not use.' The demonstration might be followed by an ultimatum to Japan, and if she then refused to surrender, by actual use of the bomb with the sanction of the United Nations.

The report made the further points that America could preserve neither secrecy nor superiority in the nuclear weapons field for more than a few years, nor could she corner the world's uranium and thorium. (The latter is another potential source of a nuclear explosive.) In the long run she would have to depend on international agreements.

The idea of staging a warning demonstration of the bomb, or simply telling the enemy of its existence and potential, had already been discussed with some thoroughness by the Interim Committee. They knew, while the Chicago scientists did not, how very few bombs there were in the pipeline, and what heavy losses might be sustained in subduing the Japanese mainland by invasion. They doubted whether the Japanese would surrender as a result merely of a warning. They also perceived various technical difficulties. For example, a demonstration bomb or one dropped after a warning might be a dud; an aircraft carrying the bomb might be shot down; or if a warning mentioned a particular target, the Japanese might move American war prisoners into the area. The Franck report therefore failed to shake the Interim Committee's recommendation.

Nor did all the scientists agree with the report – far from it. When a hundred and fifty of them were polled at the Met. Lab. on 12 July four days before the Trinity test, 15 per cent favoured full military use and 46 per cent 'a military demonstration in Japan to be followed by renewed opportunity for surrender'. Exactly how the second alternative was interpreted by those questioned is uncertain (might some have taken it to mean a demonstration without loss of life?) but it seems fair to assume that at least half the Met. Lab. scientists thought the bomb should be used in combat against Japan. At other sites the proportion would probably have been higher.

Military preparations to use the bombs, which had been started in 1944, went ahead steadily. Targets were chosen: big cities with military bases and war industries. The first list included Kyoto, once the capital of Japan, a historic city of great religious significance to the Japanese, but at the same time an ideal target for assessing the destructive power of the bomb.

Groves tried hard to avoid showing the list to Secretary of War Stimson, maintaining that it was a military matter for the Chief of Staff to decide, but Stimson insisted on seeing it and ordered the omission of Kyoto. He felt that the destruction of Kyoto would be a wanton act, reaping a harvest of bitterness after the war. Groves tried to get Kyoto reinstated during Stimson's absence at the 'Big Three' (Churchill, Stalin, and Truman) conference at Potsdam in July, but Stimson's deputy in Washington cabled Potsdam and the Minister reaffirmed his veto, which now had the President's strong support.

It was during the Potsdam conference that Truman was informed of the phenomenal success of the Trinity test of the first nuclear explosive device. Groves's report reached him on 21 July. Churchill read it the next day and told Stimson, 'Now I know what happened to Truman yesterday. I couldn't understand it. When he got to the meeting after having read the report, he was a changed man. He told the Russians just where they got on and off and generally bossed the whole meeting.'

Stalin was only told that the US had a new and unusually powerful weapon. He remarked casually that he hoped they would make good use of it, giving no sign that he had any inkling of what it actually was, despite the information passed by Fuchs and others to the USSR. A poker face presumably.

On 24 July Truman issued orders for the dropping of an atom bomb on Japan as soon after 3 August as weather permitted. On 26 July Japan was given a warning, not mentioning atomic weapons, but threatening 'prompt and utter destruction' if she failed to surrender. Some influential Japanese wanted peace, but on 28 July the Premier announced the rejection of the ultimatum.

The nuclear explosives and other bomb parts were waiting on the island of Tinian, seventeen hundred miles from Hiroshima, which was number one on the list of targets. There had been some anxiety about enemy action or accidents during transport of the materials across the Pacific, especially in the case of the planes carrying the plutonium, but everything had arrived safely. The hazards had been real enough; the cruiser *Indianapolis*, which had brought most of the 'Little Boy' bomb, was sunk by a Japanese submarine a few days later.

Little Boy was dropped on Hiroshima from a B-29 bomber accompanied by two observer planes at 09.15 (Tinian time) on 6 August. Such a small group of planes was deemed unlikely to attract an enemy attack, which could have been something of a disaster, and in fact only one enemy fighter was observed. The B-29 saw the flash of the explosion followed by 'two slaps on the plane' – the direct shock wave and the reflected wave from the ground – and then saw the huge cloud rising to 35,000 feet. Not until they were 363 miles away did the cloud disappear from sight. Reconnaissance the next day showed that 60 per cent of the city had been destroyed. The final death toll is for various reasons uncertain; the municipality of Hiroshima gave the UN Secretary-General a figure of 140,000 in 1976, but much lower and much higher figures have also been quoted.

'The important result, and the one we sought,' wrote Groves, 'was that it brought home to the Japanese leaders the utter hopelessness of their position.' Truman reinforced this when he broadcast the news to the world.

The Japanese flew two of their leading nuclear scientists over Hiroshima: Nishina on 8 August, and Arakatsu on 10 August. The pattern of destruction and the presence of radiation convinced both of them that an atom bomb was indeed responsible. Their 'Physics Colloquium' in 1943 had gravely miscalculated what America could do.

The military in the US considered it important to drop a second bomb as soon as possible, before the enemy could recover their balance, and to induce fears of bombs on one city after another. A second Little Boy would not be available for months, so it must be a Fat Man, a plutonium bomb. The date depended on having enough plutonium, and was ultimately brought forward to 9 August. The primary target was Kokura, the secondary, Nagasaki.

Weather conditions were not good, and something went wrong with the rendezvous between the B-29 bomber and the observer planes, in one of which was William Penney, a British subject, a future Director of the Atomic Weapons Research Establishment at Aldermaston in the UK. Kokura, when the B-29 arrived, was obscured by haze. After making three ineffective runs over the city the B-29 headed for Nagasaki, where a hole in the clouds appeared at the last moment, enabling the crew to aim visually instead of depending on radar, which would have been against their orders. By now they were very short of fuel, and feared they might have to ditch the plane, but they just made it to Okinawa, without even enough in the tanks to taxi from the runway.

Owing to the configuration of the land, damage and casualties were rather smaller than at Hiroshima, even though the energy released was somewhat larger; 44 per cent of the city was destroyed and there were about half as many deaths as at Hiroshima.

A bald account of the military operations and their consequences appears cold and callous. Those who planned them may seem heartless, and so too may the scientists who created the weapons. What must be borne in mind is the continuing loss of life inflicted by a suicidally determined enemy, the stories emanating from the Japanese prison camps, and the consequent overwhelming sense of urgency to finish the war.

Judgement must also be tempered by the fact that the atom bomb raids were not exceptional in terms of numbers killed and areas devastated, compared with the 'conventional' air raids that had been made on Hamburg and other German cities, and on Tokyo, where 83,000 died in an incendiary attack in March 1945. The difference, of course, was that a single bomb had achieved these results in a few seconds. The horror was heightened by the unknown in the form of nuclear radiation and fall-out, causing casualties of a new and gruesome kind, though not broadly speaking more terrible than those produced by 'conventional' means.

After the Nagasaki raid the Americans parachuted recording instruments over the city. Attached to three of these was a message to Sagane, the Japanese nuclear physicist who had worked at Berkeley before the war, 'from three of your former scientific colleagues during your stay in the United States'. These were Luis W. Alvarez, Robert Serber, and Philip Morrison, who were on Tinian at the time. They seem to have acted on their own initiative, and did not append their names to the message, but Alvarez wrote it out by hand, presumably so that Sagane could recognize the handwriting and be assured of its authenticity. The last paragraph read in part:

> We implore you to confirm these facts [that atom bombs had been used] to your leaders . . . As scientists we deplore the use to which a beautiful discovery has been put, but we can assure you that unless Japan surrenders at once, this rain of atomic bombs will increase manyfold in fury.

Even after Nagasaki and the further blow of the USSR's entry into the war against them, the Japanese Army still refused to give in. The war was nevertheless ended by the personal intervention of Emperor Hirohito on 14 August. He was possibly risking his life at the hands of military fanatics, who tried to stage a coup in Tokyo in order to continue the hopeless struggle. It certainly seems to be the case that the second atom bomb was necessary to ensure surrender.

The world was awestruck by the news of the bombs, thankful that the war was over, but aghast at the demonstrated power of the new weapons. To nearly everyone outside the Manhattan Project the atom bombs had come as a total surprise, even to those who were aware that they were possible in principle. Knowing of the leakages to the USSR, we may easily underrate the astonishing achievement of keeping so large an activity so secret for so long, especially from the Germans, who only knew rather vaguely of an Allied nuclear project and imagined it to be still at an early stage.

The Chicago scientists' prediction that the dropping of the bombs would usher in a new era was echoed all round the world. The mood was captured by Stimson at a Press conference, when he said, 'Great events have happened. The world is changed and it is time for sober thought.'

His further words, 'The focus of the problem does not lie in the atom. It resides in the heart of man,' also met with a wide response. For a short period the shock of the bombs on Japan seemed to open people's minds to basic truths. The *Saturday Review of Literature*, for example, had an editorial on the bomb entitled 'Modern Man is Obsolete', implying that whatever was done politically, on a deeper level man himself and his motivations must change. The words of Canon B. H. Streeter, the great Oxford scholar, in the 1930s came to life with a new immediacy: 'a nation [or a world] that has grown up intellectually must grow up morally or perish.' Too few knew how to convert such words into action, and the vision faded. But it was the right vision.

Some reacted very differently. There was a German prisoner of war, a fervent Nazi, who felt thrilled and exalted by the thought of the mighty power over men's lives that was packed into the atom bombs. He was envious of the airmen who had flown the planes that had carried them. 'They must have felt like gods,' he said.

To many of the scientists it was a searing experience to realize what they had helped to create. Often their feelings became stronger as time went by and the thought of the suffering in Japan sank in. During the war they had wrestled with neutron chains, barriers, racetracks, plutonium, implosion, initiators. Now they were face to face with the end-product of the technology: wounds, burns, radiation sickness, destruction, death.

'The atomic bomb is so terrible a weapon that war is now impossible,' Oppenheimer is reported to have said, foreshadowing the theory of the deterrent. Fear of nuclear retaliation does indeed seem to have been a significant factor in averting war between the super-powers.

Two who felt a particularly direct responsibility for the atom bomb were known personally to the author: Frisch, one of the authors of the memorandum that resurrected the British nuclear project in 1940, and Bohr, the doyen of atomic and nuclear science. I knew both in Copenhagen before the war, Frisch as an ingenious and enthusiastic experimenter, and Bohr as a joyous explorer of the fundamentals of the material universe. When I met Frisch again after the war, all the zest seemed to have been drained out of him, while my strong impression of Bohr was of a man carrying a heavy burden.

To the German nuclear scientists interned in Britain the news of Hiroshima was devastating. They were totally unprepared, believing an atom bomb to be still a very distant proposition. The British Officer in charge, Major T. H. Rittner, heard the news on the BBC at 6 p.m. and went to tell Hahn, one of the men who had, so to speak, lit the fuse through the discovery of fission. 'I could not believe it,' said Hahn, 'but the Major emphasized that it was no newspaperman's story but an official announcement by the President of the United States. I was almost unnerved by the thought of the great new misery it meant, but glad that it was not Germans but the Anglo-American allies who had made and used this new instrument of war.' He told Rittner that he had once contemplated suicide when he first saw that fission might lead to an atom bomb.

The other internees missed Hahn at supper, and Wirtz went to look for him. He arrived in Rittner's office just as the news was being repeated in the BBC's next bulletin. Together Hahn and Wirtz then told the others. Amid the uproar that followed Heisenberg at first energetically denied that the new weapon could really be an atom bomb. The word 'atomic' could, after all, apply to a number of things, and there had been no mention of uranium. What Heisenberg could not explain away, however, was the statement that the explosion was equivalent to that of twenty thousand tons of TNT.

Further information came in the main news at 9 p.m. including a reference to uranium and to the vast size of the American project. The Germans could doubt no longer. They were overwhelmed, stunned. They saw that, far from leading the world, they had been totally outstripped. Rosy ideas of continuing their work under Allied auspices were brusquely shattered.

One of the younger Germans, Horst Korsching, made a pertinent comment: '[The atom bomb] shows at any rate that the Americans are capable of real co-operation on a tremendous scale. That would have been impossible in Germany. Each said the other was unimportant.'

It has sometimes been suggested that the Germans' reactions were due to their ignorance of the possibility of an atom bomb. Captured German documents show, however, that the essentials were clear to them. Their amazement was rather due to their insight into the nature of the mighty task successfully performed by the Americans.*

In Japan there were two reactions: a determination never to make, own, or use nuclear weapons, and, on the part of those who knew about the Japanese wartime project, a policy of silence. American scientists visiting the country in the wake of the bombs failed to breach the curtain of silence; they were under instructions to treat their Japanese opposite numbers gently, so they did little probing. Japan acquired the image of an innocent victim. Only recently has it become apparent that her scientists and her military, like those in other countries, would have made the bomb if they could.

Hiroshima and Nagasaki gave away the Americans' principal secret, which was simply the fact that atom bombs can be made, that the theory actually works. The world now knew that there was no insuperable obstacle to the making of a weapon. Any nation with a corps of competent scientists could put them on to atom-bomb work in the expectation of a successful outcome.

* The Germans' reactions to the news of Hiroshima were picked up by a concealed microphone, but only Groves's excerpts from an English translation are available. Some of the German internees have claimed that, taken out of context, these give a misleading impression.

13 During the Cold War

The discovery of fission occupies a unique place in the history of science. No other single discovery has had such dramatic consequences in so short a time. Within four years it led to the first man-made nuclear reactor, and within three more to the atom bomb.

At Groves's request, Henry D. Smyth, who had been on the original S-1 Committee, prepared a detailed, 144-page, semi-technical account of these achievements, and this was published on 12 August 1945, only three days after Nagasaki. One of Groves's motives was to make it clear to the Manhattan Project scientists how far they could go in open discussion. Bush and Conant also wanted the report issued, and they convinced President Truman that it was needed to prevent the circulation of 'reckless and excited' versions of what had been done.

Some felt that the Smyth report revealed too much, but, as the columnist Drew Pearson said, it was no more possible to retract it than 'to pull an egg back into a chicken'. Szilard estimated that the report took the rest of the world to the point the Americans had reached in the autumn of 1942. It gave other countries a gateway into the atomic age. In the USSR there was a first printing of no fewer than thirty thousand copies of a Russian translation.

The Smyth report was followed over the next few years by the publication, mainly by the Americans, but also by the British, the Canadians, and the French, of numerous scientific and technical reports on the work carried out during the war, though these did not include the technology of nuclear explosives or of the weapons themselves.

While other countries were digesting the published material and starting to plan their nuclear programmes, the Manhattan Project in the US was running down as staff left for the universities and for industry. For a period there was confusion, with no master plan any more and uncertainty about the future of the nuclear sites. Groves knew what he wanted: to keep everything under military control. The scientists, however, fought and won the battle to create a civilian body, the US Atomic Energy Commission, to take over.

The transfer was made at the end of 1946, and the Army did not even retain Los Alamos.

An issue in the political battle was the blowing up of the Japanese cyclotrons in the autumn of 1945, on orders emanating from Groves's office. The object, of course, was to deprive Japan of anything that might conceivably help her to make an atom bomb. The scientists, however, saw the cyclotrons as pure research tools, with only very limited relevance to weapons work. If the Army did not understand that, they argued, then the Army was not competent to run the American peacetime project.

Civilian control, however, did not mean no more American bombs. It was generally taken for granted that the US must build up her nuclear arsenal. The wartime sites therefore survived, and here the scientists found plenty to do. They had had to put many interesting topics aside under the pressures of war, and now they had a chance to take them up. Among these were a wide variety of ideas for nuclear reactors, from which a commercial nuclear power programme was eventually to emerge.

For other countries the immediate possiblity was to set up a nuclear research centre. Some had sufficient scientific know-how to build their own small reactors for experimental work, and others were later able to obtain them mainly from the US. Even without a research reactor there were the applications of radio-isotopes and radiation, for instance in agriculture and medicine, which did not call for large resources, and could usefully be taken up almost anywhere.

Many people at this time had dreams of cheap and abundant nuclear power for mankind, and the highly industrialized nations especially regarded a research centre as a step towards a power programme. The centre would provide a pool of experts and in due course a base for technical planning, and meanwhile there was plenty of preparatory work, as well as pure scientific research.

The actual realization of nuclear power was still felt to be some way into the future. The Smyth report had guardedly estimated that a start might be made in ten years, but only for specialized applications. In fact nuclear power was to be used to propel an American submarine ten years later and to produce electricity on a commercial scale in the UK eleven years later. Thereafter there was to be a major expansion throughout the world.

During the intervening period, when nuclear weapons still dominated the scene, there were a series of shocks. On 6 September 1945 a cypher clerk at the Soviet Embassy in Ottawa, Igor

Gouzenko, defected and revealed the existence of a large Russian spy network in North America, which included a senior British physicist at the Montreal laboratory, Alan Nunn May. Nunn May was one of those who had been enlisted for the Communist cause while a student at Cambridge in the 1930s, but he kept rather quiet about it and grew into an inconspicuous academic. This made him the sort of man the Russians could use, and there are indications that they earmarked him as a potential spy quite early on, and then 'developed' him shortly before the Trinity test in 1945. He responded willingly and gave them nuclear secrets and materials, saying later, 'I felt this was a contribution I could make to the safety of mankind.'

Nunn May and likewise Fuchs generally refused the material rewards they were offered. They gave the USSR valuable secrets because Communism had claimed their allegiance to the point where solemn oaths cease to be binding.

Four years after the Gouzenko affair there was another major shock, a nuclear explosion in the USSR on 29 August 1949. It came as a total surprise to the West. The US had not anticipated losing her monopoly so soon, while the UK had expected to be the next after the US to make an atom bomb.

We now know that the Russian scientists achieved a chain reaction on 25 December 1946, while plutonium production seems to have started in the autumn of 1948. The Russian time-schedule was thus remarkably close to the American, but four years behind.

It was also more or less what America's own experts had predicted. At the end of the war, Bethe at Los Alamos had estimated that it would take the Russians from three to six years to develop a weapon, and others had given similar figures. The 1949 explosion should therefore not have been a surprise. What perhaps lulled people into complacency was the fact that year by year went by with no news of significant Soviet nuclear activity; it did not seem to occur to them that this was due to tight security, not lack of action.

A few months after the Russian nuclear explosion came the arrest of Klaus Fuchs on 2 February 1950, as described later in this chapter. He was a far more damaging 'atomic spy' than Nunn May, and had handed over an immense quantity of highly sensitive information over a period of years. Espionage was therefore seized on as an explanation of the Russians' rapid progress in the weapons field. Fuchs was said to have given away 'the secret of the atom bomb', as if 'the secret' could be compared with a password or the combination of a safe. In fact, the basic theory of the bomb was common knowledge, and Hiroshima and Nagasaki were proof of its validity. Beyond that there was a great deal of detailed technology, which the USSR was well able to develop for herself, though without Fuchs's information she would probably have taken a year or two longer.

What may actually have helped her most was to know from Fuchs as early as mid-1942 that the UK took an atom bomb seriously and, in 1943, that the US was mounting a major effort. This may explain why the Russians got their programme going in February 1943 in the midst of a desperate war situation. Without her own project, competently manned, all the atomic spies in the world would have been of no avail to her.

The only other country developing a weapon in these early years was the UK. The foundations of the British programme were laid at the Montreal laboratory, where Cockcroft had taken over in 1944 from Halban, who had proved difficult in more ways than one. Under Cockcroft the staff rose to about a hundred, of several different nationalities, morale was restored, and co-operation was re-established on a limited basis with the Americans.

Cockcroft had definite ideas about the shape of the future project in the UK, clearer probably than the British Government, and directed the preparations accordingly. He assumed that the first objective would be nuclear weapons, with nuclear power as a longer-term aim. General nuclear research would be needed all along, and radio-isotope work could readily be added to the programme.

The starting-point would be an experimental establishment, and this the Government set up early in 1946 at Harwell, on the site of a former Royal Air Force station on the Berkshire Downs. Known as the Atomic Energy Research Establishment, it was initially an all-purpose nuclear centre, but as the programme developed many of its functions were hived off to new establishments.

Under Cockcroft's light-reined leadership, the mainly young yet experienced staff returning from North America had to get Harwell going without American assistance. The US had adopted an isolationist nuclear policy after the war, and the Nunn May case, which broke at this time, did nothing to improve Anglo-American relations. Undeterred, Harwell had two research reactors and a large cyclotron in operation within a year or two.

Only after Harwell had been in existence for some months did the Government actually decide to manufacture atom bombs, and therefore to implement the plans that Cockcroft and his team had had in mind all along. It meant building reactors to produce plutonium, a chemical plant to separate it from the spent fuel (both at Windscale – nowadays known as Sellafield – in Cumbria) and a weapons establishment (at Aldermaston in Berkshire) to make the bombs themselves. These were followed by a diffusion plant (at Capenhurst in Cheshire) to manufacture the alternative nuclear explosive, ^{235}U.

In charge on the industrial side was Christopher Hinton, an engineer who had run ordnance factories during the war. The weapons side was entrusted to Penney, who had been at Los Alamos. With Cockcroft they formed an outstanding trio, each excellently suited for his challenging, pioneering task. Each could have had an easier or better paid life elsewhere, but chose instead to put his whole capacity into creating the new organizations needed. Between them they gave Britain a leading position in the nuclear field in the 1950s.

Close to the heart of all the early research and planning was Fuchs, who had been put in charge of the Theoretical Physics Division at Harwell on his return from Los Alamos in 1946. He even said of himself, 'I *am* Harwell.' He was still unsuspected, but Henry Arnold, the Security Officer, wondered what sort of a man he really was.

Arnold took an unorthodox line in his job. He was especially on the look-out for ideologically motivated individuals, who, he reasoned, would stand out in some way from the crowd. Fuchs was certainly an unusual type with, ironically, an obsessive concern about security. Arnold made a point of getting to know him.

Fuchs by this time was becoming increasingly disillusioned with the USSR and its post-war policies. Moreover, whereas when he first came to England his personal contacts had been almost exclusively left-wing, his work under Government auspices had brought him in touch with people of many different kinds. What he saw in some of them – 'a deep-rooted firmness which enables them to lead a decent way of life' – eventually transformed his thinking. Of himself, he said that he at last realized that there are 'certain standards of moral behaviour you cannot disregard'.

While Fuchs was fighting his inner battles, a tip-off came from the FBI in America in the summer of 1949 that atom-bomb information had been passed to the USSR, probably by a British scientist. Of a score of possibles, Arnold thought of Fuchs; but there was little hard evidence, little that would stand up in a British court of law, unless Fuchs himself provided it.

That Arnold and his colleague William Skardon induced Fuchs to confess and to co-operate fully says much for their skill and sensitivity for, unknown to them, Fuchs mistakenly imagined he faced the death penalty.

Fuchs's arrest was an ugly surprise to his colleagues. Arnold told the author that one of them travelled post-haste from Scotland to offer his help, saying 'I could not believe the accusations.' 'You will, when you hear the evidence,' replied Fuchs laconically. Another, a senior man, told Arnold, 'Even after hearing all the evidence I can hardly bring myself to believe he really did it.' Today the power of ideology to subvert is, regrettably, more familiar. What remains rare is release from that power, such as Fuchs experienced when his motivating beliefs changed – when he repented, to use the good old word.

If the change in Fuchs had been better understood at the time, he might conceivably have been treated differently. The Attorney-General, who prosecuted, saw his confession merely as 'a curious phenomenon characteristic of those queer psychological processes which some adherents of the Communist Party seem to go through'. Fuchs's naturalization was revoked, even though this meant that his highly capable brain would be available to the East rather than the West after he had served his sentence.

The first Russian nuclear explosion, the Fuchs case and, later in 1950, the disappearance from his senior post at Harwell of Pontecorvo, one of Fermi's original team in Rome, and his reappearance in the USSR, gave added urgency to the military side of the British project. The first British test explosion took place on 3 October 1952 in the Monte Bello Islands, fifty miles off the coast of Australia, making Britain the third nuclear weapons state.

The fourth was to be France. Charles de Gaulle had visited Ottawa in July 1944, and while he was there, Jules Guéron, a French scientist at the Montreal laboratory, secretly informed him about the atom bomb. After the liberation of Paris the following month, others of the Montreal Frenchmen came over to Europe and began to put Joliot in the picture. Groves, aware of Joliot's Communism, was much disturbed, but there was little he could do.

De Gaulle set up the Commissariat à l'Énergie Atomique (CEA) by an ordinance of 18 October 1945, and he said later that his aim was to enable France to make her own nuclear weapons. Soon after, however, he departed from the political scene, leaving Joliot free, at least for a time, to assert his own policy. The CEA's interests, he claimed, were purely peaceful; above all, its purpose was to provide France, weak in indigenous energy sources, with nuclear power. In fact, even if Joliot had been of a mind to undertake the making of nuclear weapons, it would have been beyond France's capabilities in the late 1940s.

The difficulties faced by the CEA in their war-torn country were indeed mountainous. One of their first objectives was a heavy water reactor. They were lucky to have sixteen tons of uranium compounds, seven which had been sent to Morocco in 1940 and hidden there, and nine which were found in a railway siding at Le Havre, where they had lain unidentified and disregarded all through the German occupation. The heavy water came from Norway. The CEA's original plan was for a reactor producing a significant amount of heat, so as to gain experience of cooling systems and to make a substantial amount of plutonium, but too little was known about the behaviour of the different reactor materials when hot; the scientists had to be content with the low-energy device. It was sometimes called the *F*rench *L*ow *O*utput *P*ile, until Joliot discovered what the English word 'flop' means.

Work started in mid-1947 at top priority under Kowarski, who had returned from Montreal the year before. The prestige of the CEA and possibly its future funding were at stake in meeting the deadline of the end of 1948, so there was a great sense of relief and triumph when the reactor became critical during the morning of 15 December.

In Britain and the US there was some consternation about the French success, lest Joliot reveal reactor secrets to the USSR. (In fact there was little the Russians did not already know.) Joliot himself roundly asserted that no honest Frenchman, Communist or otherwise, would ever hand his nation's secrets to a foreign power. For this he was publicly denounced by the Party; Jacques Duclos, its secretary, said that a progressive 'has two fatherlands, his own and the Soviet Union'.

Joliot might have resigned from the Communist Party or been expelled, but soon afterwards the Communists started an international peace campaign and wanted his oratory and the prestige of his name. He devoted more and more of his time to propaganda, and this was to the detriment of his leadership of the CEA. Finally he announced at the Communist Party's National Congress in

1950 that if asked to make nuclear weapons he would refuse, and a declaration to the same effect was circulated for signature among CEA staff.

This amounted to defiance of the Government's authority, as Joliot himself was very well aware, besides greatly embarrassing them in negotiations with the Americans. On 26 April 1950 the Prime Minister told Joliot he was dismissed.

For a year or two the CEA was in the doldrums, but then it began growing rapidly, with a programme that at that stage ran along somewhat similar lines to the British. Even with Joliot out of the way there was, however, no commitment to make nuclear weapons until de Gaulle returned to power in 1958, and the first French weapons test, in the Sahara, did not come until February 1960.

Well before this, both the US and the USSR had tested the

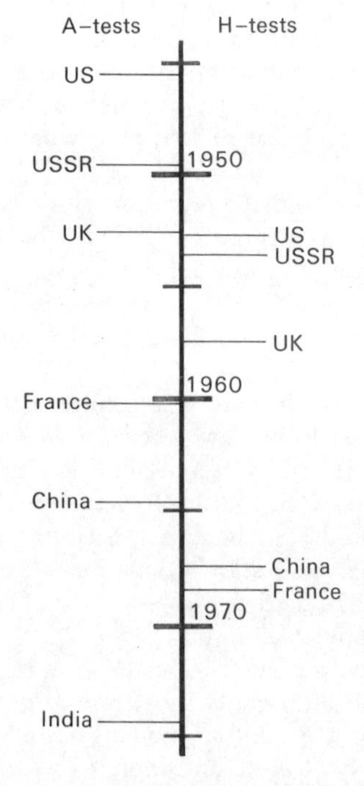

Fig. 15. Dates of first test explosions by countries indicated.
(No other countries have carried out nuclear explosions.)

hydrogen or thermonuclear bomb, the super. The US only took the decision to continue its development after widespread debate in Government circles between those who felt the safety of the country depended on it, and those who felt it was too murderous a weapon to contemplate. Among the scientists Teller was one of the chief protagonists for the super, Oppenheimer one of the chief against.

Oppenheimer's opposition to the super was to cost him dear. It was one of the principal issues during the hearings in 1954 to determine whether he was a security risk. By this time he had long since left Los Alamos. The formal charges against him were concerned chiefly with his Communist associations, but the emphasis at the hearings was on the advice he had tendered on nuclear weapons, the argument being that he had deliberately promoted the interests of the USSR. There was, however, no proof that he had done anything more than express his sincere opinion. Teller, who disagreed strongly with Oppenheimer's views, did not doubt his loyalty, though he doubted his 'wisdom and judgment'. The outcome of the hearings was the revoking of Oppenheimer's security clearance.

The decision to go ahead with the super was spurred on by the first Soviet bomb test in 1949, and was announced by Truman on 31 January 1950.

As was pointed out earlier, a fission explosion is needed to produce the enormous temperatures necessary for a fusion explosion. It is comparatively easy to use a large fission explosion to detonate a small fusion explosion. In certain circumstances the 'booster principle' may then operate, in which the neutrons from fusion increase the efficiency of the fission explosion. The Americans tested a device of this kind on 24 May 1951 at Eniwetok atoll in the Pacific.

It is very much more difficult to make a true super in which a large mass of thermonuclear fuel is ignited by means of a relatively small fission explosion. This the Americans first achieved on 1 November 1952, again at Eniwetok. The explosion, known as 'Mike', was a thousand times as powerful as the one over Hiroshima. The thermonuclear explosive in the Mike device was, however, a liquefied hydrogen isotope, liquid deuterium, which would be quite unsuitable in a weapon. It was not until the spring of 1954 that a device in which the liquid was replaced by a solid (lithium deuteride) was successfully tested by the Americans.

The Soviet Union exploded its first thermonuclear device on

12 August 1953. It was a very much smaller explosion than Mike, but it did use lithium deuteride, enabling the Russians to claim that they were ahead in the development of 'real hydrogen weapons'. Their second test on 23 November 1955 was more nearly comparable with the American 1954 test, though somewhat smaller.

Behind the Russian success lay the work of Andrei Sakharov, now known the world over for his courageous fight for human rights. He is reported to have solved several crucial theoretical problems in the making of a super, as well as to have suggested lithium deuteride for the 1953 test, and he has been dubbed 'the father of the Russian hydrogen bomb'. In 1953, at only 32, he became the youngest-ever full member of the Russian Academy of Sciences and a Hero of Soviet Labour. It was a matter of pride to his compatriots that he was entirely a Russian product, trained exclusively in the Soviet Union.

In the early 1950s he was in the main a technical man doing a brilliant technical job, but even then he read what he could get hold of on wider issues, and was impressed by Bohr's pleas for a more open world. Bohr had first expressed these in 1944 and they became his main preoccupation for the rest of his life. On 9 June 1950 he addressed a long and carefully worded Open Letter to the United Nations on the subject. To busy politicians during the cold war it must have seemed vague and impractical, but Bohr's ideas struck an answering chord in Sakharov's mind and were no doubt a factor in the remarkable development of his own thinking.

It was Sakharov's work on the super that led him to take his first steps in dissidence, protected to start with by his great reputation. Anxious about radioactive fall-out, he campaigned in 1957 for a cessation of nuclear weapons tests. Other scientific issues then began to engage his concern, for instance the bogus genetics of Trofim Lysenko that Stalin had promoted. In 1966 he finally ceased to identify himself with the Soviet 'establishment' and two years later his thoughtful book *Progress, Co-existence and Intellectual Freedom* was published in the West. His independence was by now intolerable to the regime, and one day when he arrived at the top secret scientific institute where he worked, he was refused admission. It was a hard blow, but it enabled him to spend his time succouring the victims of the regime and trying to transform the way the citizens of the USSR think and act.

Britain and France followed the US and the USSR in developing the super, and mainland China also entered the nuclear arms race, with first an atom bomb and then a hydrogen bomb. The only other country to explode a nuclear device is India, but it was not a bomb and she claims that her intentions are purely pacific.

The world was introduced to the atomic age through nuclear weapons, and weapons dominated the nuclear scene in the early post-war years. Nuclear power, its other main component, followed about ten years behind and, in the US, USSR, and the UK, developed out of the military programme. The Canadian and French nuclear power programmes also owed a lot to the Manhattan Project, though mainly indirectly through the Montreal laboratory.

No nation has taken the opposite step from nuclear power to nuclear weapons. Any country deciding to make atom bombs is indeed unlikely to use its nuclear power plants for the purpose, because power reactors are ill-suited to the task.

14 Energy for the World

The atomic or nuclear age is the age in which Man learned how to tap the energy of the atomic nucleus. He is able to do it explosively in bombs or at a steady, controllable rate in nuclear reactors, when it can be used to produce electric power. For bombs he can exploit the fission of large, heavy nuclei and for still larger bombs, the fusion of small, light nuclei. For electric power he can use fission but not so far fusion. Electricity from nuclear fusion is still a long way into the future, and it is not even certain that it is a practical proposition at all.

A full history of nuclear power would be beyond the scope of this book, but an outline will be given to indicate how the pioneering work described in the earlier chapters has been applied. It is needed, too, to give a proper balance between the civil and the military applications.

The earliest use of a nuclear reactor to produce electricity is believed to have been in December 1951 in the US. This was on an experimental basis. The first reactor designed specifically for power production, though only for demonstration purposes, was a small one at Obninsk near Moscow, which started operating in June 1954. Industrial-scale production of electricity from nuclear reactors first began in 1956 in the UK. The Queen connected a nuclear generator at Calder Hall in Cumbria to the national grid on 17 October 1956.

Power production required the harnessing of the heat from a reactor instead of throwing it away, which is what had been done during the war at the plutonium production reactors in the US. The heat in a power reactor is used to raise steam (Fig. 17), and from that stage on, a nuclear power station is like a conventional one with turbines and other machines. The basic principle involved is simple, but the technology poses novel and difficult problems, partly on acount of the high level of radiation inside the reactors.

The Calder Hall reactors in the UK still had plutonium production as their main purpose, electricity being a useful by-product, but even before they started to operate, the Government

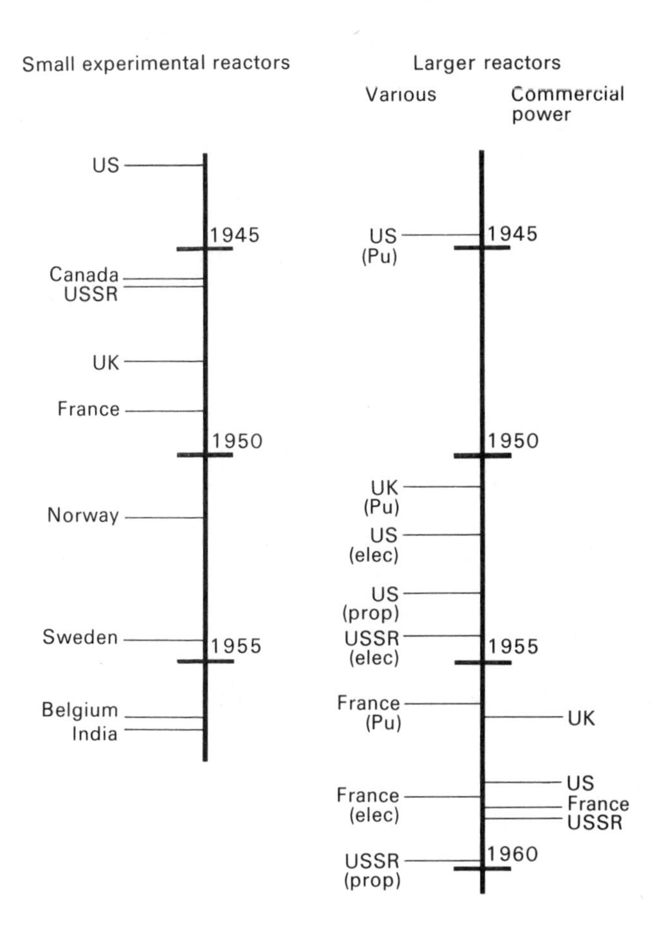

Fig. 16. Dates of the first reactors.

On the left are the dates of the earliest reactors in the countries indicated. On the right are the dates at which the countries concerned first operated larger reactors for:

(Pu)	plutonium production
(elec)	small-scale electricity production
(prop)	ship propulsion

and commercial power production. (The dates are in most cases those of achieving criticality.)

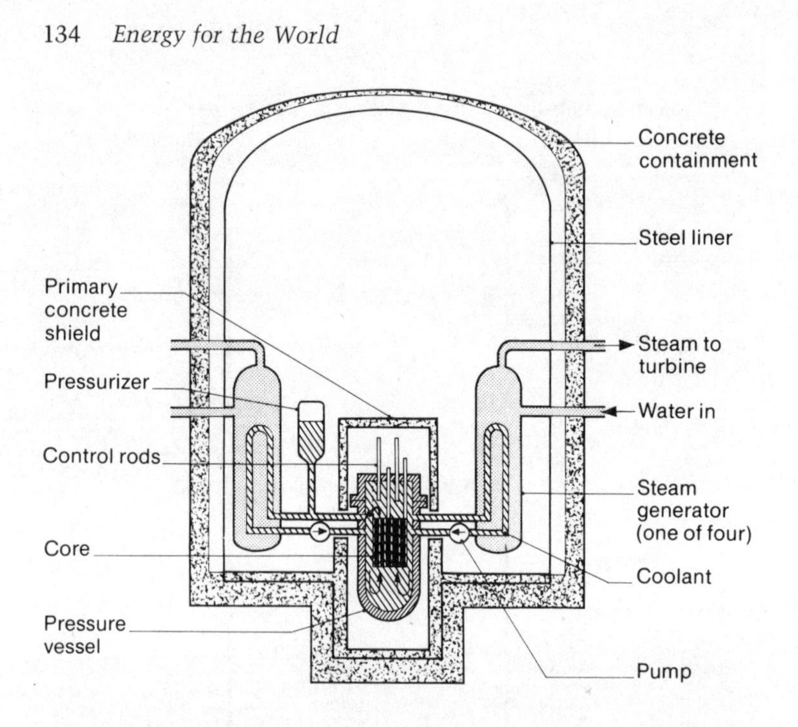

Fig. 17. The commonest type of nuclear power reactor – the pressurized water reactor.

had, in February 1955, announced a purely civil nuclear programme, and this was trebled in size in a further announcement two years later. Eighteen of the so-called Magnox reactors were built under this programme, of which the first began to generate electricity in 1962 and the last in 1971. By that time 10 per cent of the electricity in the UK was nuclear, and it was possible to boast that Britain had generated more nuclear electricity than the rest of the world put together.

Britain had forged ahead from an acute sense of an energy need. The early post-war years were dogged by energy shortages, and nuclear power promised an answer. The necessary foundation had already been laid by Cockcroft, Hinton, and their colleagues. France with less coal had an even greater need, and followed Britain as fast as her post-war recovery permitted.

The Americans, with an abundance of oil and coal, were slower off the mark. For them it was mainly a matter of choosing the cheapest method of producing electricity, and it was not until the early 1960s that the balance of costs pointed in favour of nuclear power. Once she got going, the US, not surprisingly, rapidly outstripped Britain.

Meanwhile many other countries were active, and by the end of 1983, UN information showed that there were 297 power reactors in commission in 25 countries, capable of providing just over 170 gigawatts of electricity. This is more than the *total* electric power used by Britain, France, and West Germany combined. Moreover, the plants currently under construction will more than double the output of nuclear power during the next few years and extend it to five more countries.

In some countries a fairly high percentage of the electricity is now nuclear. In Finland, France, and Sweden around 40 per cent of the electrical capacity is nuclear; in Belgium, Bulgaria, and Switzerland the figure is around 30 per cent; and in several others it is in the 10 to 20 per cent range.

By far the largest number of nuclear power reactors are of the water-cooled types developed by the Americans. Four countries, Canada, France, the UK, and the USSR, have designed and built commercial power reactors of other types and have exported a few of these. The Canadian and the main Russian types, as well as a British experimental type, are again water-cooled, but with special features. On the other hand, gas-cooled reactors were selected for the first commercial programmes in Britain and France. These offered the only chance of a quick start in the 1950s, and a very successful start it was. The eighteen Magnox reactors in the UK have been running safely and for the most part reliably for up to twenty years, and all are likely to exceed their design life. A similar statement can be made about the five Magnox reactors in France.

Magnox reactors, however, are not competitive economically with types developed later, and the American water-cooled designs, which became available just when large-scale ordering began, were able to capture most of the market. The French switched over, and only the British continued to build gas-cooled reactors on a commercial basis, having developed the Advanced Gas-cooled Reactor as a successor to Magnox; but even in Britain water reactors are likely to be started in the near future.

All the reactors so far mentioned burn only about one per cent of the uranium obtained from the ore (much of the ^{235}U *plus* a little of the ^{238}U). There is, however, a way of extracting the energy from a much higher proportion of the material. It is done by converting the ^{238}U atoms into plutonium, and burning that as a fuel. This enables us to use at least half the uranium, so increasing the energy yield fiftyfold or more.

The technology required involves the Fast Breeder Reactor (FBR), so called because it uses fast neutrons and 'breeds' plutonium from ^{238}U. Britain and France, the countries furthest ahead in this field, are at the stage of testing the first large-scale FBR prototypes. Their successful operation demonstrates the basic feasibility of the technology, and FBRs, with their associated fuel fabrication and chemical plants, could become the world's principal power producers in the next century. There is, however, a long way to go before FBRs can make a major contribution.

Nuclear power programmes, including FBR development, depend on perception of energy needs, as events during the 1970s illustrate. In 1973 there was a major oil crisis; the flow of oil from the Middle East was cut back, and the price was suddenly raised fourfold. This caused a ferment of concern about world energy resources, with the spectre of a disastrous shortfall, perhaps as early as the end of the century, bringing in its train grave shortages of electricity, transport, manufactured goods, food, and employment. Nuclear power was seen as a means of relieving the pressure. 'For heaven's sake get on with your nuclear programmes and leave us some oil,' certain Third World spokesmen are reported to have told Western delegates at a World Power Conference in the late 1970s.

The 1970s indeed saw a major expansion of nuclear power throughout the world. Soon, however, the 1973 oil price rise produced an opposite effect. It set off a world economic recession, which eased the pressure on energy resources, leading to an oil glut and an excess of electrical generating capacity. With falling demand, as well as growing environmental concern, the market for new reactors became sluggish, though France still pressed firmly forward.

The long-term dangers nevertheless remain, and if a world energy shortage develops it may take a long time to correct it, considering that ten years is commonly required to plan for and build a large modern power station, whether its fuel is oil, coal, or uranium.

Even during the present recession we are using up our oil reserves faster than we are adding to them by fresh discoveries. Barring major surprises, the world's oil and also its natural gas will begin to run out in the next few decades. This will leave just two

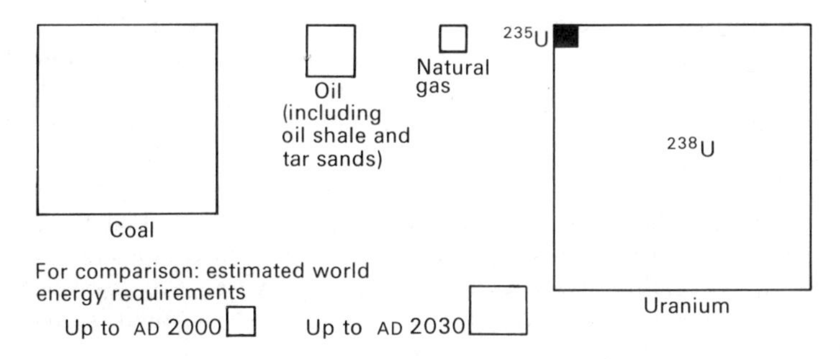

Fig. 18. Estimates of the world's available energy resources.

Uncertainties in the estimates may affect the prospective dates at which resources run short, but will probably not change the broad picture. It will be seen that, of the resources shown, only coal and ^{238}U are large enough to meet the next century's needs. Other energy resources are mentioned in the text, and the potential of some of them is relatively enormous, but their large-scale exploitation is very speculative.

energy sources that are commensurate with mankind's needs and can be exploited by existing technology: coal and uranium (Fig. 18). It is to be expected therefore that many countries will move towards a coal/uranium mix, in proportions depending on their circumstances.

The extraction of energy from uranium seems indeed to have become possible just in time, when a new energy source is needed. To some, including the author, this is evidence of God's provision for humanity.

Estimates of the quantity of uranium in the form of ores likely to be recoverable for use amount to something in the region of ten million tons. If burned in reactors of the main types in use today, this would constitute an energy resource comparable in magnitude to that of oil or natural gas – a valuable addition to our total supplies, but only capable of staving off the threatened shortages for a couple of decades at most. If burned in FBRs, however, the period would be extended from decades to centuries.

There is also a vast store of uranium in the oceans, amounting to about four thousand million tons. Potentially this is an enormous energy resource, but the uranium is exceedingly dilute, and so far it has not proved practicable to recover it, though there is continuing research on the problem.

From the point of view of next century's energy needs, the case for utilizing uranium is very strong, especially if FBRs are employed. On the other side of the argument are safety questions. These have been an issue for the nuclear industry ever since reactor design began in the Second World War, and public concern has been growing since the mid-1970s.

While reactors cannot blow up like atom bombs, designers have always recognized that they might sometimes get out of hand, and they have provided increasingly elaborate precautions against all kinds of malfunctions and imagined accidents. The measures taken have generally proved very satisfactory. Nevertheless there have been accidents including the major one at Three Mile Island in the US on 28 March 1979.

The true significance of this accident has often been missed. What it demonstrates is the effectiveness of the built-in safety systems, which protected the public in spite of an unfortunate and unlikely combination of equipment failures and operator errors.* The really serious consequences were financial, because the protective measures did not save the reactor itself, and psychological, because Three Mile Island has hardened anti-nuclear attitudes. People ask: might there be a much worse accident? Yet even when extreme improbabilities are included in the analysis – say, an aircraft crashing on a reactor that has got out of hand – the *hypothetical* consequences in terms of death and disablement would generally be less than the *actual* consequences of recent non-nuclear accidents, such as two Jumbo jets colliding or a hydroelectric dam giving way. A sense of perspective is important in our approach to these questions.

A further safety issue concerns radioactive wastes. These are an inevitable accompaniment of nuclear power. The most highly radioactive waste products result from the fission process itself, the quantities being roughly proportional to the amount of power generated. Their toxic character was noted in the Frisch and Peierls report in 1940; the Manhattan Project had to safeguard its staff from their effects, creating what was almost a new science, that of radiation protection; and the nuclear industry today is well aware of the potential hazards.

Unlike industrial slag heaps, the wastes are not of great bulk, and the usual practice hitherto has been to store them at nuclear sites,

* It has been calculated that the individual most at risk in the neighbourhood of Three Mile Island – who has actually been identified – received a radiation dose that affected his health about as much as smoking five to ten cigarettes would have done. (Not five to ten per day – five to ten in all.) Everybody else received far less.

for instance in tanks or silos, where they can be kept under surveillance. As an interim measure this has been broadly satisfactory despite occasional small leaks, and it can if necessary be continued for several decades.

For the long term, more permanent methods are being developed. It is often said that the disposal of nuclear wastes, especially those with high levels of radioactivity, is an unsolved problem, but this is only true in the sense that it is a complicated problem which still requires much detailed study. There are no inherent difficulties of such a kind as to cast doubt on the evolution of a radiologically acceptable scheme in due course; indeed we may expect a range of possible schemes to choose from. The general principle will be to provide a series of barriers between the waste form and Man. These will be the waste form itself, of glass or some other highly resistant material; the waste container, made of a long-lasting material such as stainless steel; perhaps an absorbent packing round the container, in case of water ingress; and finally the geological features of the waste repository. Leakage of radioactivity through all these barriers will be very slow, and preliminary calculations indicate that the resultant radiation doses to Man will be minute.

To the author, who has worked in this field for a number of years, the management of radioactive wastes seems a problem – or rather a series of problems, since there are various kinds of wastes – of similar difficulty to many that are tackled successfully by modern industry.

One further aspect of the energy scene must be mentioned, namely the role of the so-called 'alternative' energy sources. In some quarters it is hoped that these might make possible the early phasing out of nuclear power. The 'alternatives' are:

- Solar energy in its various forms.
- Geothermal energy – the natural heat of the Earth.
- Tidal energy.

Their importance lies especially in the fact that they constitute energy income; the supply is in most cases renewed from day to day. Fossil fuels (coal, oil, etc.) and uranium, by contrast, are capital resources; once used, they are gone. Ultimately, in a few centuries' time, it is possible that the 'alternatives' will be the only available energy sources left, unless a long shot such as nuclear fusion comes off.

The chief hope lies with solar energy. Geothermal and tidal

energy can only be harnessed on a significant scale in favoured localities, such as Iceland for the former and the Rance estuary in Brittany for the latter. In a global perspective they are of minor importance.

Solar energy, on the other hand, is a vast, world-wide resource. It originates in the nuclear fusion reactions in the sun, which are similar to those in an exploding hydrogen bomb. The amount of solar energy falling on the Earth is equal to the output of a hundred million large power stations, and if we could make use of a mere hundredth of one per cent of this, our energy worries would be over.

The problem is to collect it, because it is very thinly spread. In northern Europe the average amount on each square metre is about the same as would be obtained from a 100-watt light bulb, and surprisingly the rate even in the Sahara is only about twice as much. Solar collectors have to cover large areas, and their sheer size makes them expensive.

Fortunately, Nature transforms solar energy for us in various ways. The sun's heat sets the air in motion and evaporates water, giving wind-power, wave-power, and water-power. The sun's light is absorbed by plants, helping to produce wood and other forms of burnable vegetation. These different sources have supplied mankind with limited amounts of energy in the past. As regards large-scale applications, there is at present only one, hydro-electricity; it provides about 2 per cent of the energy used by the world, and could give substantially more.

Particularly since the 1973 oil crisis, a number of countries have attempted to develop the 'alternatives' on a bigger scale and in new ways. The principal future possibilities are wind generators, many times more powerful than the old-style windmills, generators powered by ocean waves, and fuel obtained from vegetation ('biomass'), for instance by fermentation. So far, however, research and development have not produced an obvious 'winner', a new technology that Governments can rely on to supply large amounts of energy for their people, or that entrepreneurs are glad to take up in a big way.

Research into the 'alternatives' must continue, because they are energy income. If we use all the available energy capital, they will still be there, if only we can discover how to exploit them on the scale required. Conservation of energy resources is also of obvious importance.

Meanwhile coal and uranium can give us a long breathing space.

They can meet the world's energy needs for a few centuries while we work out the next stage in the progress of mankind.

This will call for much more than technological development. The sudden shock of Hiroshima and Nagasaki awakened people briefly to the need to deal with human motives. Will the slow dwindling of our energy capital turn our minds again in this direction, with greater persistence and greater determination to find answers?

Appendix: Some Salient Facts about the Nucleus of the Atom

We can think of the *atom* as a minute ball. It consists of a cloud of tiny particles called *electrons*, and at the centre, very much more diminutive still, is a *nucleus*. Though so small, the nucleus is thousands of times as heavy as an electron and carries a positive electric charge, while the electrons are negatively charged.

The nucleus is composed of two sorts of particles, known as *protons* and *neutrons*, which are roughly equally heavy. Each proton carries a positive charge, equal but opposite to that of the electron, while the neutron has no charge at all, i.e. it is electrically neutral. The protons and neutrons are tightly bound together by a special force which operates inside atomic nuclei.

The smallest and simplest nucleus is that of the hydrogen atom, which consists of a single proton. Round it, in the neutral hydrogen atom, is a single electron, whose negative charge balances the positive charge of the proton.

There is also *heavy hydrogen* or *deuterium*, in which the nucleus contains a neutron as well as a proton. This makes the nucleus about twice as heavy, but does not change the nuclear charge or the number of electrons (one) required to make a neutral atom.

Hydrogen and deuterium provide the simplest example of *isotopes*. By this we mean that their nuclei have the same charge but different masses,* or to express the same thing in another way, the same number of protons (one) but different numbers of neutrons (nought and one respectively). To take a more complicated example, all uranium nuclei have 92 protons, while those of the two principal uranium isotopes have 143 and 146 neutrons respectively, making totals of 235 and 238 particles in all; they are designated by the symbols ^{235}U and ^{238}U. All the chemical elements exist in a number of isotopic forms.

If two atoms are isotopic, their outer parts, their electron clouds,

* We speak of mass rather than weight, because the weight of an object depends on where it is. A man, for instance, weighs less on the Moon than on the Earth, but his mass – the amount of matter he contains – remains the same.

contain the same number of electrons (e.g. one in the case of the neutral hydrogen atom, 92 in the case of neutral uranium) and are indeed virtually identical. Now it is the electron clouds that very largely determine the behaviour of an atom; when atoms meet, in chemical reaction for instance, it is primarily the electron clouds that come into contact and interact with one another. Isotopes are therefore exceedingly similar in most of their behaviour, and once mixed they are very difficult to separate. Major differences only appear in phenomena that involve the properties of the nuclei rather than the electron clouds.

The mass is one such property, and it can be exploited to separate isotopes. The *mass spectrometer* is a device which achieves this on a small scale by sorting out heavier from lighter atoms by means of electric and magnetic fields. A modified form of the same principle was applied during the 1939–45 war to separate ^{235}U from ^{238}U on a large scale.

Some nuclei are stable, others unstable. The unstable ones change into more stable nuclei in the process known as *radioactivity*. Thus ^{238}U nuclei undergo a long series of such changes, ultimately turning into stable lead nuclei, with the radium and polonium isotopes discovered by the Curies among the intermediate steps. The unstable species disappear in the process, and are said to 'decay'. When a species X decays to a species Y, we commonly speak of X as the parent and Y as the daughter.

Nuclei of ^{235}U undergo a similar but different series of changes. Though isotopic with ^{238}U, and therefore almost indistinguishable from it chemically, ^{235}U differs in its radioactivity.

In radioactive decay, the nucleus emits energetic radiations of several kinds, through which the phenomenon was originally discovered. These radiations have many remarkable properties, among them the power to penetrate matter to various extents, to fog a photographic plate, and to kill living tissue, including cancers. They include high-speed *alpha-particles* (the nuclei of helium atoms, each consisting of two protons and two neutrons); electrons and their opposite numbers, *positrons* (with the same mass as the electron, but a positive charge); and *gamma-rays* (like x-rays, but generally more energetic and penetrating).

In the decay process, the positive electric charge on the nucleus usually changes, and this means that one chemical element has been transformed into another, i.e. that a *transmutation* has taken place. Uranium, for example, ultimately transmutes to lead.

Transmutations can also be produced artificially. This was first achieved with the aid of the alpha-particles from certain radioactive substances, which were used to bombard suitable materials. Large electrical machines called particle *accelerators* (popularly 'atom smashers'), especially those known as *cyclotrons*, have now replaced radioactive preparations as sources of the high-speed particles required. Neutrons are also effective, and need not be accelerated; slow neutrons as well as fast can penetrate through the electron clouds to the nuclei and interact with them.

Quite often these transmutation processes produce radioactive species. The first example was the production of radioactive phosphorus by the action of alpha-particles on aluminium. In such cases we speak of *artificial radioactivity* to distinguish it from the radioactivity of elements such as uranium, which are found in Nature.

Fission is a special type of transmutation, in which a large nucleus divides into two main fragments along with a small number of neutrons. The main fragments are of somewhat similar though not usually equal size, and they fly apart with greater energy. Fission was originally discovered in studies of the action of slow neutrons on uranium. Although fission is usually initiated by slow neutrons or other such agents, *spontaneous fission* also occurs: the nucleus divides up without any external stimulus.

Since neutrons both produce and are produced by fission, *chain reactions* of fissions sustained by neutrons are possible. This permits the fission of very large numbers of nuclei, and hence the release of substantial amounts of energy, either explosively in bombs or at a controlled rate in power plants. In the latter case the neutrons are often slowed down by means of *moderators* such as graphite.

Nuclear energy can also be released by the *fusion* of very small nuclei. Two deuterium nuclei, for example, can unite to make a helium nucleus. The energy of the sun and the stars comes mainly from such processes. To carry out fusion on a useful scale terrestrially requires the production of temperatures comparable with those in stellar interiors, and is of course of considerable difficulty.

Further Reading

Similar ground to that of the present book is covered in:

R. W. Clark, *The Greatest Power on Earth*, Sidgwick and Jackson, London, 1980.

This is a larger, more journalistic book, containing much interesting detail. The only other comprehensive account of which the author is aware is in German; it can be strongly recommended to those who read the language:

*J. Herbig, *Kettenreaktion: das Drama der Atomphysiker*, Deutscher Taschebuch Verlag, München, 1979.

Mention may also be made of:

M. Gowing, *The Development of Atomic Energy: Chronology of Events 1939–1978*, UK Atomic Energy Authority, London, 1979.

Other books deal with specific aspects of the subject. Some are primarily historical, while others are biographical or autobiographical. In the former category are:

The American Project

H. D. Smyth, *Atomic Energy*, US Government Printing Office, 1945.

*R. G. Hewlett, O. E. Anderson, *A History of the United States Atomic Energy Commission, vol. 1: The New World*, Pennsylvania State University Press, 1962.

*R. G. Hewlett, F. Duncan, ibid., *vol. 2: Atomic Shield*, 1969.

*S. Groueff, *Manhattan Project: the Untold Story of the Making of the Bomb*, Little, Brown & Co., Boston, 1967.

H. York, *The Advisors: Oppenheimer, Teller and the Super Bomb*, Freeman, San Francisco, 1976.

The first of these is the original official report, and the next two constitute the detailed and well-documented official history. Groueff's book is a very readable popular account. York's book covers the nuclear weapons race between the US and the USSR.

The British Project

M. Gowing, *Britain and Atomic Energy 1939–1945*, Macmillan, London, 1964, and *Independence and Deterrence: Britain and Atomic Energy 1945–1952, vol. 1: Policy Making, vol. 2: Policy Execution*, Macmillan, London, 1974.

These books form the first parts of the scholarly and readable official history.

The French Project

B. Goldschmidt, *L'Aventure Atomique*, Fayard, Paris, 1962.
*S. R. Weart, *Scientists in Power*, Harvard University Press, 1979.

Goldschmidt's book is an account by one of the scientists involved. Weart's is fuller and very well researched.

The German Project

S. A. Goudsmit, *Alsos*, Henry Schuman, New York, 1947.
*D. Irving, *The Virus House: Germany's Atomic Research and Allied Counter-measures*, William Kimber, London, 1967.

The first of these is a fascinating account of the US mission sent to Europe in 1944–5 to discover the state of the German project; a few of the conclusions reached have since proved incorrect. The second is based mainly on a study of numerous official German documents; written in journalistic style, it occasionally misses the point scientifically.

The Japanese Project

C. Weiner, 'Nuclear Weapons History: Japan's Wartime Bomb Projects Revealed', *Science*, vol. 199, p. 152, 1978.

This journal article is at present the main source of information.

The Russian Project

A. Kramish, *Atomic Energy in the Soviet Union*, Stanford University Press, 1959.

This book pieces together the outlines of the history of the Russian project.

Autobiographical books have been written by several of those who knew the Manhattan Project from the inside:

A. H. Compton, *Atomic Quest: a Personal Narrative*, Oxford University Press, 1956.

L. R. Groves, *Now it can be told: the Story of the Manhattan Project*, Harper, New York, 1962.
Leona M. Libby, *The Uranium People*, Crane Russack & Charles Scribner's Sons, New York, 1979.
L. Szilard: his Version of the Facts. Selected Recollections and Correspondence, edited by S. R. Weart and G. W. Szilard, MIT Press, Cambridge, 1972.

To these may be added two biographies:

Laura Fermi, *Atoms in the Family: my Life with Enrico Fermi*, Chicago University Press, 1954.
P. Goodchild, *J. Robert Oppenheimer*, BBC Publications, London, 1980.

The latter is based on the excellent TV documentary on Oppenheimer, and contains a wealth of photographs.

There are also, of course, a number of biographies of the Curies, Rutherford, Bohr, Cockcroft, Joliot, etc. One which must specially be mentioned is:

S. Rozental (editor), *Niels Bohr: his Life and Work as seen by his Friends and Colleagues*, North-Holland, Amsterdam, 1967.

It is the source of the stories of the Bohr–Einstein argument (Chapter 1) and of Frisch's talk with Meitner (Chapter 3). Finally, dealing with Fuchs, Nunn May, and Pontecorvo, there is:

A. Moorehead, *The Traitors*, Hamish Hamilton, London, 1952.

There are many more books besides those mentioned, and asterisks have been inserted above to indicate where further titles may especially be found.

For information on energy and nuclear reactors the reader is referred to:

J. Ramage, *Energy: a Guidebook*, Oxford University Press, 1983, and the suggestions it makes for further reading.

There is also a brief but comprehensive report by the author:

H. A. C. McKay, *World Energy Resources*, AERE-R 8856, 1977.

Sources of Important Documents

Szilard's patent application (Chapter 4). *The Collected Works of Leo Szilard. Scientific Papers*, edited by B. T. Feld and G. W. Szilard, MIT Press, Cambridge, 1972.

Einstein's letter to Roosevelt (Chapter 5). R. W. Clark, *Einstein*, Hodder, London, 1973.

The Frisch and Peierls memorandum and the MAUD reports (Chapter 6). M. Gowing, *Britain and Atomic Energy 1939–1945*, Macmillan, London, 1964.

General Farrell's report on the Trinity test (Chapter 10). L. R. Groves, *Now it can be told: the Story of the Manhattan Project*, Harper, New York, 1962.

The Franck report (Chapter 12). A. K. Smith, *A Peril and a Hope: the Scientists' Movement in America 1945–47*, University of Chicago Press, 1965.

Bohr's Open Letter to the United Nations (Chapter 13). S. Rozental (editor), *Niels Bohr: his Life and Work as seen by his Friends and Colleagues*, North-Holland, Amsterdam, 1967.

Index

OPUS

General Editors
Keith Thomas
Alan Ryan
Peter Medawar

OPUS books provide concise, original, and authoritative introductions to a wide range of subjects in the humanities and sciences. They are written by experts for the general reader as well as for students.

Most of the titles listed below are only available in paperback editions; some, however, are available in both hardback and paperback, and a few in hardback only.

Further details of OPUS books, and complete lists of Oxford Paperbacks, including the World's Classics, Twentieth-Century Classics, Oxford Shakespeare, Oxford Authors, Past Masters, as well as OPUS series, can be obtained from the General Publicity Department, Oxford University Press, Walton Street, Oxford OX2 6DP.

In the USA, complete lists are available from the Paperbacks Marketing Manager, Oxford University Press, 200 Madison Avenue, New York, NY 10016.

Architecture

The Shapes of Structure
Heather Martienssen

Business

The Way People Work
Job Satisfaction and the Challenge of Change
Christine Howarth

Economics

The Economics of Money
A. C. L. Day

History

The Industrial Revolution, 1760–1830
T. S. Ashton

Karl Marx
His Life and Environment
Isaiah Berlin

Early Modern France, 1560–1715
Robin Briggs

Modern Spain, 1875–1980
Raymond Carr

The Workshop of the World
British Economic History from 1820 to 1880
J. D. Chambers

English Towns in Transition, 1500–1700
Peter Clark and Paul Slack

The Economy of England, 1450–1750
Donald C. Coleman

The Impact of English Towns, 1700–1800
P. J. Corfield

The Russian Revolution, 1917–1932
Sheila Fitzpatrick

War in European History
Michael Howard

England and Ireland since 1800
Patrick O'Farrell

Louis XIV
David Ogg

The First World War
Keith Robbins

The French Revolution
J. M. Roberts

The Voice of the Past
Oral History
Paul Thompson

Town, City, and Nation
England 1850–1914
P. J. Waller

Britain in the Age of Economic Management
An Economic History since 1939
John Wright

Language

The English Language
Robert Burchfield
forthcoming

Law

Law and Modern Society
P. S. Atiyah

Introduction to English Law
Revised Edition
William Geldart

English Courts of Law
H. G. Hanbury and
D. C. M. Yardley

Literature

The Modern American Novel
Malcolm Bradbury

This Stage-Play World
English Literature and its Background, 1580–1625
Julia Briggs

Medieval Writers and their Work
English Literature and its Background, 1100–1500
J. A. Burrow

Romantics, Rebels and Reactionaries
English Literature and its Background, 1760–1830
Marilyn Butler

Ancient Greek Literature
Kenneth Dover and others

British Theatre since 1955
Ronald Hayman

Modern English Literature
W. W. Robson

Mathematics

What is Mathematical Logic?
J. N. Crossley and others

Medical Sciences

What is Psychotherapy?
Sidney Bloch

Man Against Disease
Preventive Medicine
J. A. Muir Gray

Philosophy

Aristotle the Philosopher
J. L. Ackrill

The Philosophy of Aristotle
D. J. Allan

The Standing of Psychoanalysis
B. A. Farrell

The Character of Mind
Colin McGinn

Moral Philosophy
D. D. Raphael

The Problems of Philosophy
Bertrand Russell

Structuralism and Since
From Lévi-Strauss to Derrida
Edited by John Sturrock

Free Will and Responsibility
Jennifer Trusted

Ethics since 1900
Mary Warnock

Existentialism
Mary Warnock

Politics and International Affairs

Devolution
Vernon Bogdanor

Marx's Social Theory
Terrell Carver

Contemporary International Theory and the Behaviour of States
Joseph Frankel

International Relations in a Changing World
Joseph Frankel

The Life and Times of Liberal Democracy
C. B. Macpherson

The Nature of American Politics
H. G. Nicholas

English Local Government Reformed
Lord Redcliffe-Maud and
Bruce Wood

Religion

An Introduction to the Philosophy of Religion
Brian Davies

Islam
An Historical Survey
H. A. R. Gibb

What is Theology?
Maurice Wiles

Hinduism
R. C. Zaehner

Science

Science and Technology in World Development
Robin Clarke
forthcoming

The Philosophies of Science
An Introductory Survey
R. Harré

A Historical Introduction to the Philosophy of Science
J. P. Losee

The Making of the Atomic Age
Alwyn McKay

The Structure of the Universe
Jayant V. Narlikar

Violent Phenomena in the Universe
Jayant V. Narlikar

What is Ecology?
Denis F. Owen

Energy: A Guidebook
Janet Ramage

The Problems of Evolution
Mark Ridley
forthcoming

Social Sciences

Change in British Society
A. H. Halsey

Towns and Cities
Emrys Jones

Social Anthropology
Godfrey Lienhardt

Races of Africa
C. G. Seligman

Urban Planning in Rich and Poor Countries
Hugh Stretton